CORLISS
STEAM ENGINES

F. W. SHILLITTO, Jr.

Lindsay Publications Inc.

Yours truly,
Frank William Shilletto, Jr.

HAND BOOK

—OF—

CORLISS STEAM ENGINES

DESCRIBING IN A COMPREHENSIVE MANNER THE
ERECTION OF ENGINES, THE ADJUSTMENT
OF THE CORLISS VALVE GEAR, AND
THE CARE AND MANAGEMENT
OF CORLISS STEAM ENGINES.

—BY—

F. W. SHILLITTO, Jr.

THIRD EDITION.

Handbook of
Corliss
Steam Engines

by F. W. Shillitto, Jr.

Original copyright 1898
by F. W. Shillitto, Jr.
Published by
The American Industrial Publishing Co.,
Bridgeport Connecticut, 1902

Reprinted by
Lindsay Publications Inc
Bradley IL 60915

ISBN 1-55918-000-5

1 2 3 4 5 6 7 8 9 0

1988

PREFACE.

THE demand for an elementary treatise on the Corliss Engine has induced me to undertake the preparation of this volume. It is presented with no journalistic pretensions and with no ambition save the advancement and welfare of the younger members of our chosen calling. It has been the aim of the author to set forth the principles governing the subject matter in language free from unnecessary technicalities and as concisely as possible.

While a few indicator diagrams have been introduced from the author's practice for the purpose of illustrating certain points, no attempt has been made to treat at length on this subject, as the fraternity is well supplied with most admirable works on this line.

Should this volume be the means of imparting the desired knowledge to its readers the author will, indeed, be amply repaid for the effort.

THE AUTHOR.

JUNE, 1898.

WARNING

Remember that the materials and methods described here are from another era. Workers were less safety conscious then, and some methods may be downright dangerous. Be careful! Use good solid judgement in your work, and think ahead. Lindsay Publications Inc. has not tested these methods and materials and does not endorse them. Our job is merely to pass along to you information from another era. Safety is your responsibility.

Write for a complete catalog of unusual books available from:

Lindsay Publications Inc
PO Box 12
Bradley IL 60915-0012

TABLE OF CONTENTS.

PART I.

Erecting Corliss Steam Engines.

PART II.

Adjusting Corliss Valves.

ERECTING CORLISS ENGINES.

PART I.

CHAPTER I.   PREPARING FOUNDATIONS.

When a new engine is to be installed it is to be expected that the engineer in charge should be qualified to offer valuable suggestions regarding location, etc., also to perform the actual work of erection if called upon to do so.

The following explanation of the method of procedure, aside from a few suggestions regarding location, preparing the ground, etc., will apply as well to the erection of the motive power for an entirely new manufacturing plant as to an addition to a plant already in use.

We will assume that it has been decided to install a new engine to replace one which can no longer drive the manufacturing plant at its full capacity. The first thing to consider is the location. Generally speaking the engine should be located as centrally as possible as regards the distribution of power, that is, in case a long line of shafting is to be driven, it will be much better to locate about

the middle of the line, if possible, than
drive it from one end, as for a given
amount of power to be transmitted a
lighter shaft can be used in the former
position than is possible in the latter.
Of course it is not advisable to separate
the engine and boiler rooms by any
great distance if it can be avoided, but
the inevitable loss of time due to shutting down the plant long enough to remove the old engine and foundation,
build new foundations and erect the
new engine upon the site of the old one,
will usually far more than offset any
gain by having a compact plant. At
the present day, with our admirable
non-heat-conducting coverings, return
traps, steam loops, etc., it is possible
to conduct steam to considerable distances with but very trifling losses
from radiation and condensation.

There are many other points than
those enumerated which must be considered in deciding upon the location,
for every particular case has special
peculiarities.

In a new manufacturing plant there
should be very little difficulty in deciding upon the location of the motive
power, and yet it is regrettable that
there exists to-day so many examples
of short-sightedness in this respect,

such as engine rooms without cellars, with steam and water pipes running under the floor where there is barely room enough to crawl, to say nothing of doing effective work, when repairs have to be made, in such cramped quarters, and engines located right out in the main shop without any protection against dust and dirt.

Having decided upon the location for our new engine, the ground must be staked out for the foundation excavation—the drawings furnished by the engine builders giving all the required dimensions—the principal requirement being that it be dug with its longest side perpendicular to the line shaft in the factory.

The nature of the soil met with will have its effect upon the method of preparing for the foundation proper, therefore it is impossible to state a general rule governing all cases. A practical mason, experienced in this line of work, would be the most likely person to decide upon what is to be done in unusual cases, but the following has been found to meet ordinary requirements.

Carry the excavation down about twenty inches below where the bottom of where the brick-work is to begin, have its surface levelled and thoroughly tamped, keeping it quite damp while the tamping is being done. After it has been given a good, honest ramming fill in this extra depth,—a thin layer at a time—with a concrete composed of five parts of broken stone, two parts of clean, sharp sand and one part of Portland cement. As each layer is put in, tamp it down well before putting in the next layer until the required thickness is reached.

This will take time but it will be time well spent, as it must be remembered that even a poorly built engine may run well upon a good foundation, while the best engine built will not give satisfaction if set upon a poor foundation. Pay no attention to those who advocate economizing in material and use only the best.

The concrete bed should be given time to harden thoroughly before starting upon the foundation proper.

CHAPTER II.   REFERENCE LINES.

When the engine is set up its crank shaft must lie parallel with the line shaft—or jack shaft if there is one—in the factory, consequently the center line through the engine must stand at a right angle or perpendicular to this line shaft, therefore it will be necessary to bring a line into the new engine room to set the template to. Targets may be then set up in the engine room and this reference line preserved, for we shall have a use for it later.

Select two points on the line shaft as far apart as possible, and clear a space under the shaft between these two points, then caliper the shaft and see that the spots selected are of the same diameter, and if so we can go ahead, but if they are of different diameters, allowance must be made for the difference, and the points which we are to locate upon the floor, corrected accordingly, for it must be understood that it is the line through the center of the line shaft that we desire to locate. Under the points selected tack down squares of hard

wood, or better yet new sheets of tin
to carry the points.

You will now require a reliable
plumb bob. The affair usually sold
in the hardware stores, made of brass
are usually cast hollow and filled with
lead, and I have never seen one which
could be relied upon to locate a point.
Let one of these get to spinning, and
ninety-nine times in a hundred its
point will describe a circle, thus proving its center of gravity to be anywhere but directly over its point
where it should be. There are to be
found upon the market cylindrical plumb bobs, bored and turned from the solid bar, and filled
with mercury (quick silver); they are
reliable and are made by a firm with
a reputation for producing accurate
tools.

The writer some years ago, having
the difficulty mentioned above, made
an experimental plumb bob of cast
iron and tool steel (gleanings from the
scrap pile) weighing two and one half
pounds, which has given excellent results. It is illustrated in Fig. 1,
which gives the dimensions. It will
be seen that it may be used either end
up, by reversing the weight upon the

steel spindle, but it is much steadier when used as shown.

Returning to our shaft we now plumb down and locate points upon the spots prepared to receive them, as shown at a and b in Fig. 2. Now

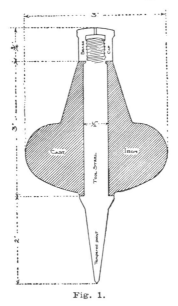

Fig. 1.

drive a fine nail half the diameter of the line to one side of point a and attach a fine braided line to it, and stretch the line through point b and fasten it to another wire nail a few feet beyond, as at c., then by tapping

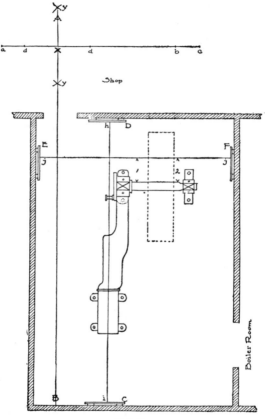

Fig. 2.

the nails, which support the line, sideways, the line may be made to exactly cut through points a and b which it is rquired to do. It will be a good idea to plumb down from the shaft again from points a foot or two inside the original points, thus proving the setting of the line. You cannot be too particular in laying out this line, because if it is out ever so little, all lines taken from it will be out accordingly.

Now select a point upon the line opposite the door opening into the engine room—as at x in the figure—lay off on the line at each side of x and, say six feet from it, points dd. For these measurements use only a light baton ten feet or more long, laying the required distance off upon it, then transfer it to the line. Long measurements made with a two foot rule or ordinary tape measure are apt to be unreliable. Drive sharp pointed wire nails, one through each end of the baton with their points projecting from the same side; this we will use as a tram and lay off from points dd, the intersecting arcs yy. These latter had better be scribed upon sheets of tin as before.

Stretch another line through points yy (where the arcs intersect) down the engine room and fasten it temporarily in this position. This line is shown at AB. The next thing to do is to set up the "targets" C D, which are to remain as permanent reference points until the work is completed and the engine running.

Get two pieces of clear pine four inches wide and two feet long, about one inch thick, also four pieces of

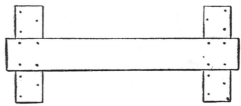

Fig. 3.

about the same width and thickness, but about a foot long, and nail them to the wall about four feet above the floor, with the middle of the length of the long piece, opposite where the center line of the engine will come, by measuring roughly from line A B., and be sure to have the top edge of the target level. The target will then appear like Fig. 3.

If the belt holes have already been located in the wall take this for a starting point; if not find where it is intended to locate the receiving pulley on the line shaft and measure back from this point to the line A B, and transfer this distance to a point upon the wall inside the engine room. The distance between the center line of the engine and the center of the band wheel having been located upon the template it is obvious that these points upon the template and the wall must be opposite each other, therefore measure back from the location of the belt wheel center, a distance equal to the distance between the engine line and belt wheel center, and locate point h upon the target D.

Measure the distance from line A B, to point h on target D (using a baton for this purpose) and going to the other end of the line A B locate point i, the same distance from A B, upon target C. The points h and i can be made permanent by making a deep knife cut or scratch upon the top edges of the targets, using a small try square to guide the blade. It will be well to put up another set of targets high enough above the first ones·to be out of reach of accident or persons bent on mischief. This is easily accomplished by plumming up from the

lower ones. The lines may be taken
up now.

If it is possible to obtain the use of
a surveyor's transit for a little while
the reference line through the engine
room may be quickly and accurately
located, the method being about as
follows: Select a point opposite the
engine room door and at this point
plumb down from the shaft to a sheet
of tin upon the floor thus locating a
point to start from, and set up the
transit with its plumb bob exactly
on this point, and level the instrument
in each direction by the aid of the adjusting screws under the frame. Select a point upon the shaft, (being
careful about the shaft diameter as
before) as far away from the transit
as possible, and suspend a plumb line
over the same side of the shaft as was
used to plumb from before, letting the
bob hang in a pail of water to bring it
to rest quickly. Now train the tele-
scope upon this plumb line, bringing
the cross hairs to bear upon it. Take
the reading of the horizontal vernier
and then swing the scope around exactly 90 degrees, (as indicated upon the
vernier) training it through the engine room door. Set up a target at B,
(fig. 2.) and the cross hairs will exactly

locate one end of line A B upon it. This being marked and a target being temporarily erected across the door-way, the other end of the line may be as readily established. This line may now be transferred to targets C D as before.

CHAPTER III.   THE TEMPLATE.

In building the foundation proper
it will be necessary to have a template, or pattern of the engine base,
with all anchor bolts accurately located thereon, to be used as a guide
for the mason to work to, so while
waiting for the completion of the preliminary work it will be advisable to
get one out, if one has not been furnished by the engine builders.

The drawing referred to in Chapter
I should give the exact location of
each bolt hole, as compared with the
center line of the engine, and the center line of the crank shaft, so we will
transfer these points on to the template, which will of course be the full
size of the engine base.

One inch boards eight inches wide
will answer for the main parts of the
template, and one inch by six inches
will be all right for cross pieces representing the engine feet, also the diagonal brace.   The drawing will also
give the distance from the center line
of the engine to the center of the fly
wheel, which should also be laid off
upon the template.   The template
when completed will appear like Fig. 4.

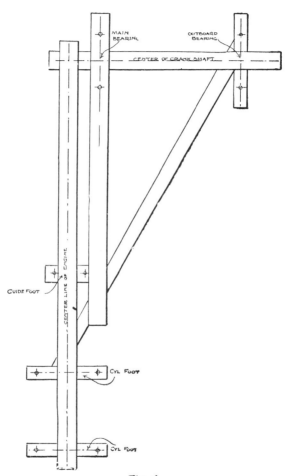

Fig. 4.

The bolt holes should be bored in the template, of such size as will fit the bolts snugly.

The anchor bolts should not be "built in" under any circumstances, owing to the difficulty in getting their length above the cap stones, or sole plates, just right, also of replacing one should it at any time be necessary to do so.

As casings for the anchor bolts,

Fig. 5.

tapering square wooden spouts of about five eighth inch stuff should be built, long enough to reach from the top of the "bottom stones" or anchor plates, to the bottom of the cap stones, and having their inside measure at the top, one and one half inches larger than the diameter of the bolt, while their bottom end may be just large enough to allow the bolt to enter freely. Their appearance will be like Fig. 5.

CHAPTER IV. FOUNDATIONS.

Undoubtedly the best material for an engine foundation is a good quality of hard brick, laid in a mortar composed of equal parts of sand and Portland cement, using a liberal supply of water upon the brick. A good plan is to lay up a course or two around the outer edge of the foundation, dividing up the enclosed space thus formed, by laying partitions across it, for convenience in working, and, being sure the outside courses are tight, pour in a supply of mortar almost as thin as is used for grouting, and lay the brick right in it bedding them well down and together. Then wash all the chinks full of mortar, before starting another thickness. This method has been used several times by the author and has given excellent satisfaction.

The concrete sub-foundation having become thoroughly seasoned we will proceed to set up the template. This may be supported by a frame work built up from the floor or it may be suspended from the ceiling above, the latter being preferable, when it can be conveniently done, owing to the extra facilities thus afforded for getting around under the template.

Stretch a line between the targets—
C D, fig. 2—through points h and i,
and draw it very tightly and whichever method of supporting the template is used, place it under the line
at a height above the concrete equal
to the depth of the foundation, and
approximately center it by plumming
down from the line onto the center
line on the template. If it is suspended, after getting the perpendiculars
up, it may be drawn either one way or
the other as is required to accurately
center it, by diagonal braces. After
it has been securely fastened any temporary support may be removed.

Suspend plumb lines through the
centers of the bolt holes in the template, which will give the proper location for the holes in the anchor plates,
or bottom stones,—which are heavy
iron plates or square cut stones with
holes drilled so that the bolts may
pass through them,—which are to be
set in after say four courses of brick-
work have been laid, leaving "pockets" directly underneath the anchor
plates, for putting the washer and nut
on the bottom ends of the bolts. After the anchor plates have been set in,
the wooden casings may be placed in position and the bolts with their top nuts

on dropped through their respective holes in the template, cases, and anchor plates, and their weight supported by blocking under their ends in the pockets.　The cases should then be adjusted so as to leave an equal space all around the bolts at their top and then nailed in this position to the template.　The brickwork will now be plain sailing until the time has arrived to set the cap stones, when it will be necessary to remove the template. The tops of these cap stones on the main portion of foundation should all lie level and in the same plane, as nearly as possible; the stone under the outboard bearing is usually eighteen or twenty inches higher than the others, the drawing giving this required data.

Figure 6 illustrates a brick foundation with iron anchor plates, extending through from side to side, and granite cap stones, which will be found to give satisfaction. We consider that heavy cast iron plates, well ribbed on their backs (top sides), with bolt holes cored, are just as reliable as, and less expensive than, cut bottom stones.

After the bond between the cap stones and the brick has thoroughly

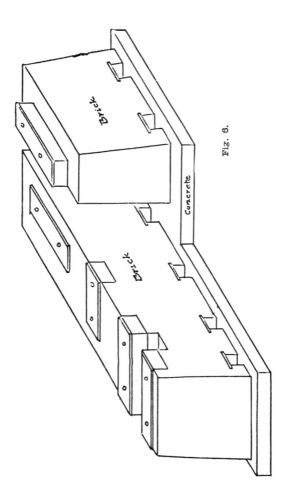

Fig. 6.

set, the tops of the stones under the cylinder, frame, and main bearing should be dressed so that they are level and their top surfaces all lie in the same plane. They should not require much dressing, for a good mason can make them lie very nearly as required without much trouble. A long straight edge, a reliable level and a good bush hammer are all the tools required for this work.

CHAPTER V. PLACING THE MAIN PARTS.

When the time comes to get in the main parts of the engine arrange it so they will come in proper order, that is the parts belonging farthest from the entrance, should come in first so as to avoid moving heavy parts around as much as possible.

You will need, for levelling the engine about twenty-four iron wedges about two inches wide, six inches long, and seven eighths of an inch thick at the large end, tapering down quite sharp at the other. These will be used between the cap stones and the engine feet.

Take the top nuts off the anchor bolts and let the bolts drop down into the pockets out of the way, then get in the half of the fly wheel that is without the key way, lower it down into the wheel pit, and chock it.

Now get in the main bearing, frame and cylinder and place them in position with wedges well entered under their feet, in the positions indicated at x in Fig. 7, and bolt these parts together, being careful to remove all for-

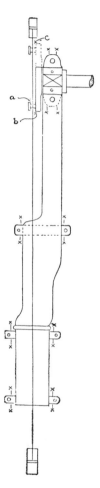

Fig. 7.

eign matter from the surfaces of the
permanent joints between the cylinder and frame, and between the frame
and main bearing. The cylinder and
guide section have been together once
in the shop and put in perfect allignment, consequently they ought to go
together again without trouble.   In
bolting them together set the nuts up
evenly, and not very tightly, all round
then finish by tightening opposite
bolts so as not to throw these parts
out of line.

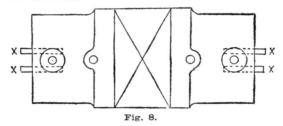

Fig. 8.

Raise the anchor bolts up through
the engine feet and put the top nuts
on loosely, leaving ample room for levelling. If any of the bolts are liable
to bind when moving the engine sideways relieve them now.

Get in the outboard bearing and set
it in position, placing wedges as
shown at x in Fig. 8, then with a
straight edge placed through both
shaft bearings level across to see if the

outboard bearing cap stone is low enough, and leave about five eighths of an inch between the foot of the bearing and the cap stone for cement filling, or if soft metal or sulphur is to be used a smaller space may be left.

Put the boxes in the shaft bearings and place the shaft in position, and put the washers and nuts on the bottom ends of all the anchor bolts.

Set up the targets E F, as in fig. 2, on a level with the engine shaft, after which we are ready to begin lining and levelling the engine.

CHAPTER VI.   LINING AND LEVELLING.

The principal requirements of lining
an engine are, that the center line
through the cylinder shall be perpendicular to the center line of the crank
shaft, and both centers must lie in the
same plane; the wearing surfaces of the
guides must be parallel to the center
line of the cylinder, and, with bored
cylindrical guides, any plane cutting
through the center line of the cylinder longitudinally, must also cut the
center line of the guides.   With V
guides it often happens that the wearing surfaces are not at equal distances
from the center line of the engine and
we could never quite understand why
they were so constructed, yet this fact
has no practical bearing upon the subject as far as this style of guide is con-
cerned.

A method of supporting the line
which has been used almost universal-
ly for years consisted of attaching it
to an upright located on the floor at
the crank end, the other end being held
by a "spider" or cross-bar in the head
end of the cylinder.   It is easily seen
that, with this arrangment, every time

the cylinder end of the engine was moved the line was carried with it which was very troublesome to the beginner. A later and much simpler way is to set up the line entirely free from the engine, and bring the engine up to the line. This is the method we shall use.

The crank shaft must be brought parallel with the line shaft in the mill, so we must establish a line in the engine room representing the axis of the line shaft. This line will be established on the targets E F in fig. 2. If there is no way of taking direct measurements for this line as by going through the doorway at one end and through the belt hole near the other end, we must apply the same method as was used to bring the line A B into the engine room and then plumb up to the targets, locating points j j, through which stretch the line.

Crowd the shaft back in its bearings —toward the cylinder—by drawing up the wedges or setting up the screws for adjusting the quarter boxes, as the case may be, wedging the shaft quite tightly.

With a light stick caliper from each end of the shaft to the line as at 1 and

2, (fig. 2).   Place one end of the stick against the shaft near one end, bring the other end of the stick up under the line and make a fine knife cut on the stick where the line crosses it. Make the other end of the shaft come up to the same relative position by swinging the outward bearing as required.   After this is done get the shaft level.   The levelling can best be done with a plumb line, as follows: Place the crank on the top quarter (the crank standing up vertically) and suspend a plumb line from above so that it hangs opposite the center of the crank-shaft and an inch or two away from the crank-pin. Measure from the end of the crank-pin to the line, then roll the crank over to the bottom quarter and measure again. These measurements must be made equal by raising or lowering the outboard bearing the desired amount.

Always watch your previous work when making a new adjustment, because in levelling a part you will most likely throw it out of line, and vice versa, therefore throughout the entire work bring up the lining and levelling together.

When you have got the shaft level and in line, tighten up the anchor bolts

holding the bearings, taking care that
the level or alignment is not impaired
by doing so.

Make two upright frames of inch
stuff, six inches wide and long enough
to come a few inches higher above the
floor, than the center line of the engine,
and put a two inch hole in each upright about on a line with the center

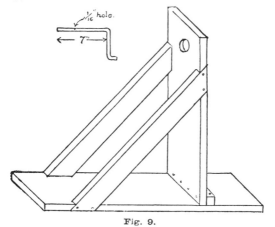

Fig. 9.

line of the cylinder, as illustrated in
Fig. 9. Set up the frames one at each
end of the engine with working room
between the engine and the uprights,
and the holes in line with the center
line of the engine, fastening them to
the floor by one end so that they may
be swung to either side for squaring

the line with the crank shaft. A piece
of three-eights inch round iron bent as
shown in Figure 9, the long arm seven
inches long with a sixteenth inch hole
drilled through it three inches from
the end, as shown, will be found very
convenient for tightening the line and
anchoring one end of it. The other
end may be fastened to a short piece
of heavy wire or light rod.

For a "line" we should recommend
very fine piano wire as it is much
stronger than any equally fine fibrous
line, and has a much nicer surface to
caliper to. Run the wire through the
cylinder and both uprights, fastening
the cylinder end of it to a short piece
of rod placed crosswise of the hole in
the upright at this end, then pass the
other end of the wire through the
small hole in the bent iron crank,
take up the slack by hand, take a few
turns around the iron crank and cut
off any surplus wire. By letting the
small crank shaft rest against the
back of the upright, and turning the
crank, the line may be drawn up very
tightly after which the crank may be
pushed around and held fast at the
back of the upright thus securely anchoring the line.

The engine line must be got square with the crank shaft, passing opposite the shaft's center, and exactly over the middle of the crank pin's length. The line may be squared by the same process as was used for levelling the shaft. Referring to Figure 7 it will be seen that the line is over the center of the crank-pin's length when the spaces a and b are equal; and it is square with the shaft when the measurements b and c are equal, (c being measured with the crank near the other center as shown dotted in). While taking these measurements be sure that all end play in the shaft is taken up by crowding it back toward the outboard bearing. This is very important and if not seen to will cause trouble.

When the line has been set as required, fasten the line supports securely and see that the line has not been moved in doing so.

Now measure up roughly to find how much the cylinder end is out horizontally and move it accordingly, when it will be found to be very nearly in its proper position, and is ready for the first levelling.

Having provided yourself with a good machinist's level about two feet

long, apply it to the bottom of the cylinder bore, and along the top of the
steam valve chambers, and get the cylinder level both ways; at the same
time bring the guide section level, by
levelling along the bottom guide; also
plumb across the finished edges of
both guides as illustrated in Figure 10.
Get these spots right, being sure that
all the wedges under the feet have
good bearings.

Our next move is to set the line

Fig. 10.

level, or parallel to the bottom guide
which we have just levelled, at the
same time keeping it in its previous
position horizontally. For this purpose make a caliper of a piece of pine,
the long arm being about as big as a
lead pencil, and with a thin semicircu-
lar base, set on edge, as shown at A
in Figure 11, the total length of the
caliper being about one-half inch

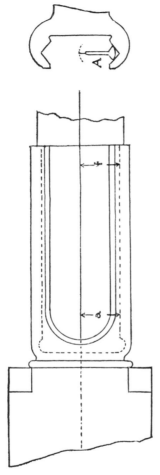

Fig. 11.

shorter than the distance from the line
to the bottom guide.   Drive a pin
straight into the end of the long arm
to adjust the caliper by. Caliper from
the lower guide to the line at points
a and f, Figure 11, as far apart as possible, and make these measurements
equal, at the same time keeping the
line cutting the center line of the
crank shaft. For example, should the
line be higher at a than at f you must
lower the cylinder end and raise the
crank end so as to keep the line opposite the crank shaft center, manipulat-
ing it so as to get the measurements
a and f equal.

It is obvious that if the guide section is level, and the line is made
parallel with it on the same plane as
the center of the crank shaft, the line
should be almost in the exact center of
the stuffing-box vertically.   After levelling the line test it once more for
squareness with the shaft, and correct
any error here.

We now have the line level, square
with the shaft, and on the same plane
as the center of the crank shaft, and
the engine is level.   What remains to
be done is to set the engine so that the
line shall be exactly centered in the
cylinder, in the center of the stuffing-

box, parallel to the guides and over the center of the guides, also level in both directions.

Make a light wooden caliper, one half inch shorter than the radius of the cylinder and stick a pin straight into the end; also make a much shorter one for the stuffing-box.

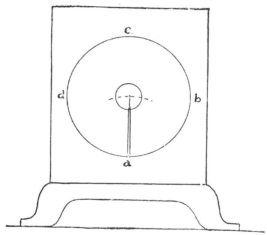

Fig. 12.

You will now perhaps find that the engine is out more horizontally than vertically, so try to correct this first. With the long caliper measure in the head end of the cylinder from points d and b, Figure 12, to the line, and move the cylinder, to the right or left, as the case may be, so as to make these meas-

urements equal.   Bring the stuffing-
box end up at the same time by simil-
arly calipering with the small caliper.

Now try the level again, and don't
try to get the last hair's breadth on the
line when the level is out, which it undoubtedly will be if the cylinder has
been moved much.   If the level is out
drive up the wedges at the required
points and caliper to the line again
both vertically and horizontally.

Fig. 13.

We will now see how the guides
stand horizontally as compared with
the line.   For this purpose make a
wooden straight edge long enough to
more than span across the edges of
both guides, drive a stout wire into
the middle of its length, and use it as
illustrated in Figure 13.   The object
is to get the edges of the guides paral-

lel with the line. Guage from each
end of the guides, and bring the frame
up as required.

Sometimes if the cylinder joint is
not carefully cleaned, a small particle
of solid matter being left adhering to
one of the surfaces, it will throw the
guides around out of line, or the same
thing may result in the bolting, or it
still may have the appearance of being out of line if one side of the frame
is a trifle lower than the other, thus
rocking it to one side. Try all these
points, and if the guides are plumb
and the frame seems to be out sideways, slack up on the apparently tight
side of the cylinder joint, tighten up
on the opposite side, then tighten up
the other side again, and most likely it
will be all right.

Having got the cylinder and guides
in line horizontally and plumb and
level, bring them up to the line vertically following the same principles as
before. When you are satisfied that
the engine as a whole is level and in
line as it stands, see that the wedges
all have an equal bearing and set up
on the anchor bolts quite snugly all
around. Try the line and level again
in all directions, because it is possible
to spring the engine down or to rock

it to one side in tightening the bolts;
this must be remedied by the wedges
and another trial made.   When tightening up the bolts quite strongly does
not disturb the level nor allignment,
you can consider the job done.

Patience, close observation, and ac-
curacy are the principle requirements
in lining an engine; without exercising
these virtues you cannot hope for suc-
cess.

The joint between the cap-stones
and the engine feet had better be made
now.   First stuff some waste around
the anchor bolts, poking it down into
the boxes an inch or two after which
poke sand in on top of it; this will keep
the filling from running into the boxes
and grouting the bolts.

If the space between the cap stone
and castings is three-eights of an inch
or less, a filling composed of seven
parts lead and one part antimony will
make a very satisfactory joint.   If
there are very thin spaces to be filled,
spray kerosene into the opening and
pour the hot metal quickly and the
space may be very easily filled.

Should the opening be half inch or
larger used best quality Portland cement, mixed clear and quite thin.

Of course provision must be made for running the filling just where it is wanted, by making a dam of sand all round each foot, with space sufficiently wide to pour the filling.

After the joints have been given sufficient time to set thoroughly—at least twenty-four hours if cement is used—tighten all the anchor bolts permanently. Two men with a six foot wrench is about right for a two-inch bolt.

CHAPTER VII. ASSEMBLING THE MOVING PARTS.

Before placing the fly-wheel in position, the shaft boxes should be scraped to a good bearing. Hoist up the shaft and slip on the eccentric, then lightly coat the surface of the journals with red lead, replace the shaft and roll it in its bearings a few times to mark the babbitt where it bears too hard. Scrape down these "high spots" and try it again, continuing to mark and scrape until the journals bear evenly in their boxes. When this is satisfactorily accomplished give the journals a coat of clean oil, put the shaft in place, adjust the bearings properly and put on the caps, taking care to plug the oil holes to keep out dirt.

Now turn the shaft so as to bring the key seat uppermost, and try the key both in the shaft and in the wheel to see that it is a proper fit, and finding it to be satisfactory, seat it well in its place in the shaft. This may seem to some to be a radical departure from the usual practice, as engine builders have been in the habit of driving the keys, which is all right with solid wheels for obvious reasons. As regards a sectional or "split" wheel, a few moments study of the situation will convince the most skeptical that it is easier, takes less time, is safer, in fact, is better every way to clamp the wheel on to a properly fitted key than it is to drive the key, especially if the key be a large one.

We have seen keys as small as one inch square that fitted the key seats in both shaft and pulley beautifully and could be seated in either with a few very light taps with a stick of wood, and, although the key seats were accurately in line (as they always should be), the key could not be driven without upsetting and throwing the wheel out of true.

Now carefully clean the wheel fit and the bore and facings of the wheel, and jack the lower half of the wheel— which is in the wheel pit—up to its place against the shaft. Sling the other half, hoist it into position and lower it into place. Put in two hub bolts diagonally opposite and draw them down solidly, then examine the holes for the rim bolts to see that they come fair when the edges of the fly wheel are true. Any holes which do

not come exactly fair should be ream-
ed true and new bolts fitted.

It is by far the best plan to shrink
the hub bolts in.   Take the other two
hub bolts and heat them evenly, in a
wood fire, to a very low heat; a red
heat barely perceptable in broad day-
light, or "black-hot," is hot enough, because if you have them red hot they
will only stretch when you put the
wrench on, thus weakening them.
Having got them to the proper heat,
put them in their holes and draw the
nuts up solid with a good stout
wrench, and the shrinkage will do the
rest.   Remove the bolts previously
put in to hold the wheel and treat
them the same way.   Now bolt the
rim and you will have a job to be de-
pended upon.

It will be advisable to set the steam-
fitters at work on the steam and exhaust piping just as soon as possible
after the engine is set, as we shall
have a use for them before the piston
and valves are in.   In the meantime
get the eccentric strap on, the rocker
arm and wrist plate stud set up, the
wrist plate on, and the eccentric and
reach rods connected, and the governor set up.

If the engine is to be run condensing do not connect the exhaust pipe to the condenser yet, but blank off this connection and first use the outboard or "free" exhaust. Bolt pieces of plank over the front ends of the valve chambers and put the bonnets on the back ends. Put the cylinder head on, and clamp a piece of board over the stuffing-box, using the gland for a strap.

Now if the piping is finished and has been tested with steam on, caution the fireman to look out for the water in the boiler, and give the pipes and cylinder a good blast of steam. Do not keep the throttle open more than a second or two, as the excessive draught of steam may cause the boilers to prime, and thus draw the water down dangerously low in the boilers. It is simply astonishing the amount of grit which may be removed from the steam pipe and engine in this manner. This pipe scale and core sand if left to itself is very apt to seriously injure the valves, piston rings, and cylinder.

The writer once took charge of a new 18 & 42x36 inch cross-compound condensing engine, which had been set up the year previous to his engagement, and had been run five days on trial. When we opened the high pres-

sure cylinder, we found conclusive
proof that this piping and engine had
not been blown out with steam before
using.   There were three very bad
grooves about one-quarter inch wide
extending the entire length of the bottom of the cylinder bore, and upon
looking further, loose scale and core
sand was found in the exhaust chest
and receiver.   Thus it will be understood that this steam scouring process is well worth the trouble.

Having repeated the blowing a few
times, at intervals, the cylinder may
be opened and wiped clean.   Then
get in the piston and piston rod,
(which in sizes up to 22 inches are
usually shipped in one piece, boxed).
Take the piston all apart, clean it
thoroughly, and examine it carefully
to see that all the parts are there and
that they fit properly, then get the piston into the cylinder.

Put in the chunk ring, packing
rings, springs, and centering screws,
and accurately center the piston in the
cylinder bore, by calipering from the
turned boss on the piston, to the counter-bore, and adjusting the screws between the spider and the chunk ring.

Put on the follower and see that
there is a good bearing or counter sink

for the heads of the bolts to seat against, also that the bolts do not bottom in the holes before they are screwed up, and set them all up hard. Remember that it is possible to put such a strain on these bolts as to cause them to break, still the one which should happen to work out, through not being screwed home, may break the cylinder head or the piston, or both.

If the piston rod is held in the piston with a nut, screw this up as solidly as possible and put, say, three good deep center punch marks between the nut and the rod, right on the thread. These will prevent the nut working off, and should it be necessary to remove it at any time, the center punch marks may be easily drilled out with a breast drill.

The piston is usually marked O or T, for the top, but if it is not, mark it so for future reference, and put the same mark on the other end of the piston rod, near the thread, so that in screwing the rod into the cross head, you may keep the piston right side up.

Get the cross-head in place and screw the piston rod into it, and set up the piston rod nut. Before we go any further with this portion we had

better adjust the cross-head in the
guides, the idea being to center the
rod with the center line of the engine,
and as we have already centered one
end of it when we centered the piston,
all that is necessary now is to get the
rod parallel to the lower guide by calipering from the guide to the rod,—as
in Figure 11, at a and f,—and raising
or lowering the cross-head through
the medium of its adjusting screws or
wedges. When you have adjusted the
bottom shoe satisfactorily, adjust the
top shoe so that there is a very slight
amount of room between it and the
top guide. Now push the cross-head
to the other end of its travel and see
that the top shoe is as free there as at
the other end, as it should be if the
guides have been properly machined.

The next thing is to locate the
"striking points" of the piston upon
the lower guide. These striking
points are lines, one near each end of
and permanently marked upon the
lower guide and denote the position of
a similar line upon some fixed point
on the cross-head when the piston is
in contact with either cylinder head.
In an engine whose piston rod is keyed
into the cross-head they are very readily located; but when the piston rod is

screwed into the cross-head, unless the exact position or depth has been marked upon the rod when they were put together in the shop, it will take a little manoeuvering to properly locate its exact position. It is evident that the connecting rod with its connections may be considered as having a fixed length, (a properly fitted rod requires no "shimming" behind the brasses), therefore we will start with the rod and locate the travel of the cross-head, by making faint "clearance" lines upon the guide, and work back from them, to locate the "striking points."

In putting on the connecting rod, key it up tightly onto either pin and see that it points fairly to the other one, thus ascertaining if the brasses have been properly fitted. Try this from both pins, and if much of an error is found here the brasses should be re-fitted.

Having the connecting rod on, place the engine on the crank end center and scribe a faint line on the cross-head and extend it across the edge of the lower guide; place the engine on the other center and scribe another line,—co-incident with that one already upon the cross-head,—upon the other end of the guide. These lines represent the travel of the cross-head, consequently the stroke of the engine.

Next measure the "inset" of the cylinder head (i. e., the depth of that part which extends into the cylinder, measured from the face of the joint), and transfer this depth to the counter-bore and mark it. Now disconnect the crank end of the connecting rod, and let it rest on blocking, or hang suspended by the tackle used to put it in place, and draw the piston up against the frame head. Cut a straight stick —a piece of seven-eights stuff two or three inches wide is just the thing— accurately to the length of the stroke of the engine, with the ends *square*, verify it by comparing it with the marks laid off on the guide, and, finding it correct lay it on its edge in the cylinder with one end up against the piston. The distance between the end of the stick and the position of the cylinder head inset as marked in the counter-bore will be the sum of the clearance for both ends. Suppose this measures five-eights of an inch, it is evident that the clearance will be five-sixteenths of an inch in each end, that is the piston should be made to travel to within five-sixteenths of an inch of each head.

Now push the cross head to the head end of its stroke as will be indicated by the marks on it and the guide being in line, and turn the piston rod into or out of the cross head as required to bring the piston five sixteenths of an inch further in the cylinder than the mark in the counter bore, and secure the rod in this position, previously seeing that the O on the rod is on top. Now draw the piston up against the frame head, when the mark on the cross-head will be found to have travelled by the one on the guides just five-sixteenths of an inch. You may now make a *permanent* line on the guide in line with that on the cross head; then push the piston up to the mark in the counterbore in the head end and the lines on the cross head and guide at this end of the stroke will be five-sixteenths apart also. Make a permanent line at this end, same as at the other. The marks nearest to the ends of the guide are the "striking points" and should be marked O as should the line on the cross head. You can now verify the work.

It is a good plan to put a prick-punch mark in the center of the O on the piston rod and another one upon

the cross head some even number of
inches from the one on the rod.   Lay
off the distance upon some finished
part of the frame for future reference
as a tram gauge, and when everything is finally adjusted locate these
tram marks, one each side of the
screwed connections of the eccentric
and carrier rods.   Should it be necessary at anytime to separate any of
these connections they may be very
easily and accurately re-adjusted by
taking up the distance laid off, upon a
pair of dividers and bringing the
marks up to this gauge.

The striking points for an engine
whose piston rod is keyed into the
cross-head are located by keying in the
rod and simply pulling the piston up
against either head (or up to the distance that the inset of the head enters
the cylinder if that head is off) and locating the marks upon the guide and
cross-head after which the connecting
rod may be put on and the amount of
clearance   ascertained   and   the   rod
lengths verified.

Give the bore of the cylinder a good
coating of cylinder oil and put on the
cylinder head to keep out the dirt.

The valves are usually shipped
each pinned to its own stem; this is

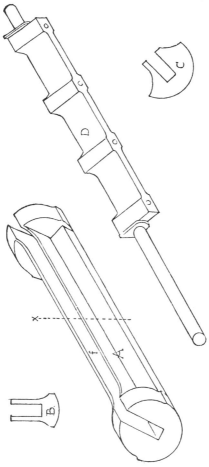

Fig. 14.

due to the fact that when a valve is being fitted to the bore of its chamber, it is turned on its own stem. Before putting them in, take out the pins and clean the valves and stems thoroughly, and examine them carefully noting the difference in shape between the steam and exhaust valves. Their general appearance is very similar, the distinguishing feature being the greater breath of the face—about one-third of its circumference—on the exhaust valve. This is required on account of the larger size of the exhaust port, also its position. In Fig. 14, A gives an idea of the general appearance of a Corliss valve in outline; B a cross section—through point x—of a steam valve, and C a cross section of an exhaust valve, through the same point. D illustrates the valve stem, usually made of phosphor-bronze, the flattened portion or "blade" being an easy fit in the slot f of the valve. Turning up the bottom edge of the blade you will find four holes about one-half inch in diameter in its edge. These holes are seatings for the short, stout spiral springs which come in the box with the valves, and when the valve and stem is put together ready for placing in its proper chamber, these springs tend to thrust the valve away from the stem, thus keeping it normally to its seat—the steam pressure acts in the same direction—and at the same time allowing it comparative freedom. The pins which held the valves and stems together must, of course, be taken out and kept out, their mission being ended practically after the valve has been turned up to fit.

You will find the valves and stems each marked consecutively from 1 to 4 corresponding to a like number stamped on the back ends of the valve chambers; these denote the chamber that each individual valve was fitted to.

Now put the valves in their proper places and put on the front bonnets— those on the valve-motion side—and bolt them fast. Push each valve snugly up against the front bonnets and try the back bonnets to see that they do not bind the valves end-ways. These points are supposed to be all right when the parts left the shop, still it is well to look into such matters and be satisfied yourself. Should you find any valve or stem that is a trifle long it, or they, must be removed and a chip turned off the back end as required to free it.

Get the valve motion and dash pots set up, during which operation no difficulty should be met with as they all have been together in the shop and properly marked. Be sure that there is no cramp or bind in any of the valve or governor rod connections, for if they are not perfectly free, they will cause trouble. Also see that the wrist plate can be moved through its extreme travel without any of the connections interfering or bringing up solidly, and the engine is all ready for valve adjustment.

PART II.

ADJUSTING CORLISS VALVES.

GEORGE H. CORLISS,

INVENTOR OF THE CORLISS STEAM ENGINE.

CHAPTER I —THE VALVE.

Before going into the details of adjusting the valves of a Corliss engine, it will be advisable to consider the construction and different functions of the common slide-valve.

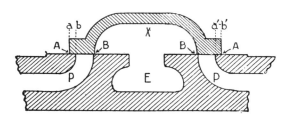

Fig. 1.

Referring to figure 1, P P are the cylinder steam ports. E is the cylinder exhaust port, and X is the exhaust cavity of the valve. The edges A.A. are the steam edges, or the edges which control the admission of steam to the cylinder and the point at which the steam is cut-off. B B are the exhaust edges, and control the opening for ex haust and the closing for compression.

In this type of valve these points are determined in the design of the engine, and are therefore unadjustable. Any change in the steam distribution would necessitate the designing of an entirely new valve, unless the desired change be very slight, when the valve may possibly be altered to meet the requirements. With the Corliss valve this would be unnecessary as will be explained at another time.

It will be seen by referring to the figure that the steam edges of the valve overlap the ports, as shown by the dotted lines a, b, and a' b'. This over-lapping is technically called "lap," and when given to a valve, as in the figure, it is for the purpose of cutting off the steam before the completion of the pis-ton stroke. The exhaust edges of the valve are "line and line" which is usual practice, yet conditions may sometimes require a small amount of inside lap to prevent a too early release.

The greatest disadvantage attending the use of the slide valve, lies in its limited ability to handle steam expan-

sively, the earliest point at which it can
be made to cut-off the steam with economy being about three-quarter stroke ;
an earlier cut-off produces a correspond-
ingly early exhaust opening for release
and an equally early exhaust closure for
compression. To put it more plainly :—
If the valve had no lap, neither steam
nor exhaust, and stood "line and line"

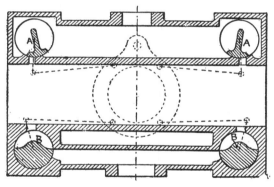

Fig. 2.

the eccentric would stand at a position
90 degrees in advance of the crank, and
the valve would then admit steam full
stroke. As lap is added for the purpose of producing an earlier cut-off the
eccentric would have to be advanced
to a greater angle, or sufficient to "take
up the lap," and have the valve in a

position to open for admission at the proper moment. It is this advancement of the eccentric which brings about the objections previously spoken of pertaining to release and compression. A too early release prevents the full realization of expansion, and over compression lessens the available net power of the engine.

In the Corliss valve gear these objections are practically eliminated.

Comparing fig. 1, with the sectional view of the Corliss cylinder and valves, fig 2, it will be seen that the four functions of admission, cut-off, release and compression, are obtained by two sets of valves in the latter, each set—one steam valve and one exhaust valve—controlling the four points for their own end of the cylinder. They may therefore be considered as the two working edges of one end of the slide valve, separated and arranged to give the greatest flexibility of adjustment, that is the Corliss steam valve, A. fig. 2, may be taken as representing the edge A. fig. 1, of the slide-valve, and the exhaust valve B, fig. 2, considered as the edge B of the slide valve. The four valves will consequently perform the same duties as the four edges of the slide-valve while possessing the extra advantages of being placed nearer the work, thus reducing clearance, and being adjustably connected to a common center of motion. This center of motion is called the "wrist plate," and its use presents the advantages of a peculiarly accelerated and retarded motion of the valves at a time to give the most beneficial results, *i. e.*, the ports are opened and closed very rapidly, and held open in such a manner as to give the least loss of pressure in admission, and the lowest back-pressure during exhaust.

CHAPTER II.—VALVE GEARS.

There is a great variety of releasing gears as applied to the Corliss engine, yet they differ only in detail and not in principle, and may, for convenience, be divided into two classes.

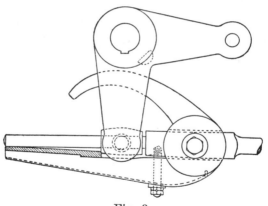

Fig. 3.

Those engines, whose valves rotate toward the center of the cylinder in admitting steam, may be considered as the first class, and include the "crab-claw

gear," Fig. 3, as originally applied by
George H. Corliss and William A.
Harris, and still used either in the
original or a modified form by several
later builders. The Reynolds-Corliss,
Philadelphia-Corliss engines, and several other makes, belong to this class

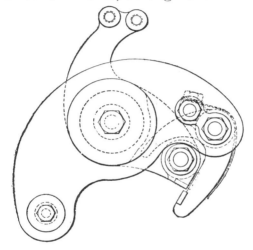

Fig. 4.

also, but are equipped with a device
known as the "half-moon gear," Fig. 4.

The second class is made up of those
engines in which the steam valves rotate toward the ends of the cylinder, or
outward, when opening for admission,
generally using a form of gear styled

the "oval arm gear," Fig. 5. To this class belong the Allis-Corliss and Hewes and Phillips-Corliss engines. There are a few builders who use the oval arm gear to rotate the steam valves toward the center of the cylinder in opening— therefore, their engines may be consid-

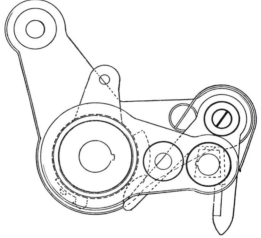

Fig. 5.

ered as being in the first class—but the gear is necessarily reversed—that is, the valve lever, or "Jim crank," hangs downward instead of standing up from the valve stem. The Hamilton-Corliss engine is a familiar illustration of this style.

CHAPTER III —SQUARING THE VALVES.

Let us now imagine before us a new 20-inch Corliss engine, set up, lined, and levelled, all parts assembled and ready for the adjustment of the valves.

The first step to be taken is technically

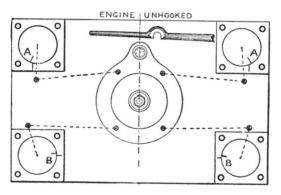

ENGINE UNHOOKED

Fig. 6.

called "squaring the valves." On removing the back bonnets of the valve chambers you will find marks on the end of each valve and on the end of each valve chamber, each of which should exactly coincide with the *work-*

ing edge of its own valve, or port, as
the case may be. It will be advisable
to inspect these points and become
thoroughly familiar with them. See
Fig. 6.

On the wrist-plate stud will be found

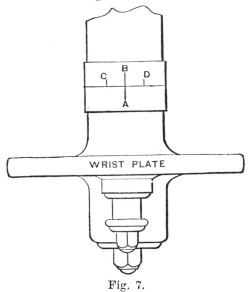

Fig. 7.

a center line, which coincides with a
similar line on the back of the hub, and
at points equi-distant on each side of
the center line of the stud there will be
found other lines, which represent the
extreme travel or oscillation of the wrist

plate in either direction when in proper adjustment. See Fig. 7, which is a top view of a wrist plate when on its center of travel, A B being the centre lines, C and D representing the extreme throw marks.

Set the wrist plate on the center and slack off the nut which holds the wrist plate on the stud, then, after interposing a piece of card board between the washer and wrist plate hub, screw up the nut hard enough to prevent the wrist plate from being accidentally moved off its center while working on the radial rods—as the connections between the valve cranks and the wrist plate are sometimes called.

Referring to the "Table of Laps and Lead," we find that a 20-inch engine requires a steam lap—*i. e.*, the distance the steam valve overlaps the port *in excess* of complete closure—of $\frac{1}{4}$ inch and an exhaust lap of $\frac{1}{16}$ inch when the wrist plate is on the center of travel, also a steam lead of $\frac{1}{32}$ inch, which, for the time being, we will not consider.

The adjustments for each end of the cylinder are obtained by lengthening or shortening the radial rods, as the conditions may require, until the lines on the steam valve—for the crab claw or half-moon gear, or any gear which opens the steam valves toward the center of the cylinder—are $\frac{1}{4}$ inch nearer the ends of the cylinder than those on the end faces of the steam valve chambers.

In any of the gears which open their steam valves outward, as the oval-arm gear, these lines should be separated the same distance in the other direction —that is, the line on the steam valve should be $\frac{1}{4}$ inch nearer the center of the cylinder than that on the chamber for the same size of cylinder.

Having made the required adjustments on the steam valves, treat the exhaust valves the same way, with the exception, of course, of the amount of lap, remembering that the *working* exhaust port is the opening between the exhaust valve chamber and the exhaust chest (see Fig. 2) and not on the port opening directly from the cylinder; therefore, the lapping of the exhaust valves will be indicated by the distance that the line on the valve is away from the line on the chamber in a direction toward the top of the cylinder or vertically. (See Fig. 6.)

There is considerable difference of opinion upon this point of exhaust lap; it formally was, and is still with some builders, the custom to give exhaust *opening* with the wrist plate central; still others place the exhaust valves "line and line," but the best practice seems to require a slight lapping of the exhaust valves when in this position.

The measurements for valve setting as given in the table are all right for ordinary practice, but in some instance they will, perhaps, require modification to fit the conditions under which the engine is to run, and considerable deviation may be made from them without seriously impairing the steam distribution. By lapping the exhaust valves more, an earlier exhaust closure will be realized, giving more compression, and at the same time a later release. It will be seen that it is not desirable to go to extremes.

The only true way after getting a new installation to work is to apply the indicator and from its readings correct any slight misadjustment that may exist, but this will be explained in another chapter.

Having carefully adjusted and fastened all connections, the valves are now "squared" and the temporary card board fastening may be removed from wrist plate and the nut tightened up.

TABLE SHOWING LAP AND LEAD OF VALVES OF CORLISS ENGINE :

Cylinder Diameter in Inches.	Wrist Plate on its Center.		Steam Lead Engine on Center
	Steam Lap.	Exhaust Lap.	
8, 10 & 12.	3-16"	1-32"	1-32"
14, 16, 18 & 20.	¼"	1-16"	1-32"
22, 24, 26, 28 & 30.	5-16"	3-32"	3-64"
32, 34 & 36,	⅜"	⅛"	1-16"

CHAPTER IV.—THE DASH-POT RODS.

The dash-pot rods must be adjusted
to the proper length ; and at this point
we must speak a word or two of caution, for should these adjustments be
incorrectly made, either the valves will

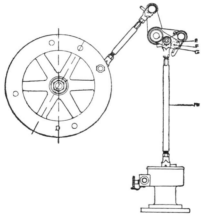

FIG. 8.

not hook up or something will be bent
or broken at the first revolution of the
engine. That is, if the rods are left too
long the closing shoulder on the releasing gear will bring up against the

hook-block before the wrist-plate has reached its extreme point of travel and either buckle the rod or break off the valve crank. Therefore, great care must be exercised at this point.

. Unhook the steam valves, allowing the dash-pot plungers to drop home, being sure that they *are* home, driving them down with a block of wood to make sure ; then carefully throw the wrist-plate over to its extreme travel and adjust the length of the dash-pot rod, H, Fig 8, so that there will be an equal space between the hook block F, and the latch steel on one side, (see G), and the hook-block and the closing shoulder on the other (as at E.)

Serve the other end of the valve gear in the same manner, and then verify these adjustments by hooking up both valves and releasing them again once or twice, and if everything is clear we are through with the valve gear for a time.

It will sometimes happen that after a new engine has been run a day or two the valves will not hook up, or may "miss" occasionally. This is due to the leathers on the dash-pot plungers becoming pliable and probably compressed a trifle, thus allowing them to drop lower and with greater freedom. When this occurs it is only necessary to carefully lengthen the dash-pot rod so that the valves will hook on, bearing in mind the point relating to clearance previously mentioned.

Too much air cushion in the dash-pot may cause the plunger to drop only partially home, thus requiring it to be pushed down by the closing shoulder on the end of the radial rod. This shoulder, by the way, is located as mentioned, in the crab-claw gear only, while in the oval arm gear, or half moon gear, it is the squared projection at the bottom of the jaw of the latch. The remedy in this case is to so regulate the amount of cushion that the plunger will drop home *rapidly*, yet without pound or jar.

Insufficient cylinder lubrication will at times have the effect of making the steam valves close slowly and also requiring them to be pushed shut, and the uninitiated may often attribute this to some derangement of the dash-pot.

CHAPTER V.—ECCENTRIC ROD, ROCKER ARM AND REACH ROD.

In determining the proper length for the eccentric rod, the proper position of the eccentric, laterally, must be found, and care taken to prevent its being moved along the shaft afterward, so as to bring it out of line either toward the main bearing or toward the fly-wheel, either of which will cause the strap to bend sidewise and give trouble by heating. To determine this position, take off the front half of the eccentric strap, and, having previously keyed up the other end of the rod tightly in position, push the back half of strap far enough back to admit of the rod being swung a trifle sidewise, as shown in Fig. 9. A little lateral movement may always be found at the strap end of the rod, enabling it to be swung sidewise probably an eighth of an inch. Take up whatever *free* play there is and note how far the

strap clears the eccentric on each side,
see a and b in the figure, place the
eccentric so that these measurements
will be equal, and mark the shaft with a

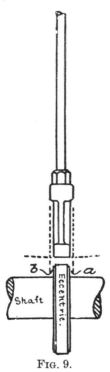

FIG. 9.

scriber at one side of the eccentric so
that this position may always be found.
The strap may be put together again
and attention given to the rocker-arm.

It is essential that the rocker-arm should oscillate equally to each side of a vertical line dropped through its cen-

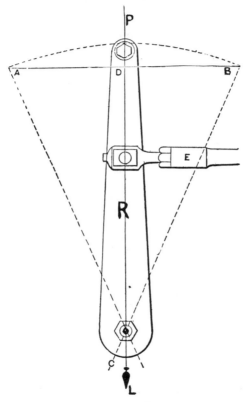

FIG. 10.

ter of support, as illustrated in Fig. 10, in which R represents the rocker-arm, or carrier-arm, as it is often called, P–L being a plumb line suspended from above in such position as to cut through the center of the reach rod stud D and then center of rock shaft O. The points A and B are its extreme travel in either direction. Rotate the eccentric around the shaft, leaving the reach rod un- hooked from the wrist-plate. Should it be found that the rocker travels far- ther toward the cylinder than toward the crank-shaft, when the eccentric is thus rotated, it is evident that the ec- centric rod is too long and it must be shortened by adjusting at E (see Fig. 10), or at the eccentric strap to an amount equal to one half the error.

Should the rocker-arm travel farther toward the crank shaft than toward the cylinder, the rod is of course too short, and the foregoing adjustments must be reversed. When the rocker-arm has been made to travel equal distances to each side of P–L, the eccentric may be partially rotated around the shaft un- til the rocker-arm stands exactly plumb once more, the reach-rod hooked on to the wrist plate, and the length of this reach-rod adjusted so that the center lines on the wrist plate hub and stud exactly coincide (see Fig. 7), care being taken that the rocker-arm is not moved off the perpendicular.

After proving these adjustments as a whole by rolling the eccentric around the shaft with everything hooked on, we are ready to center the engine and set the eccentric.

CHAPTER VI.—CENTERING THE ENGINE.

There are numerous methods of placing an engine on the dead center, a few of which will be described.

If the strap end of the connecting rod is a true surface and you have a good level, the engine may be conveniently centered by placing the level on the crank-pin strap and turning the engine so as to bring the connecting rod to a dead level at whichever end of the stroke it is desired to find the dead point.

Another method is to stretch a line parallel to the center line of the engine, running it exactly opposite the centers of the crank-shaft and the wrist-pin, or crosshead-pin, as it is frequently called, then by bringing the crank-pin center to the line the engine is on a dead center.

Still another exceedingly simple yet most accurrate way to accomplish the desired result, when the engine is constructed with an ordinary bed-plate, or

sole plate, which has been planed, is
by the use of a surface-gauge. Set up
the surface-gauge opposite the crank
and adjust the pointer to enter the center of the crank-shaft, when, by slid-
ing the surface-gauge toward whichever
dead center it is desired to find, and
then bringing the crank-pin center into
such a position that the pointer may
fairly enter it, the job is done.

The best method for general application is by "tramming" the fly-wheel,
or the disc crank, if the engine is built
with one. This method is illustrated
in Fig. 11, in which the line A–B is the
center line of the engine, and the space
between the points a and b on this
line represent the stroke of the cross-
head. Turn the engine toward the center on which you desire to place it, until the cross-head has reached a point
within an inch or two of the end of its
stroke, and then stop. Now scribe a
line across the lower portion of the
cross-head and the lower guide, this
line is represented in the figure by the
vertical line through the point c. Next
make a mark on the rim of the fly-
wheel at some exact distance from a

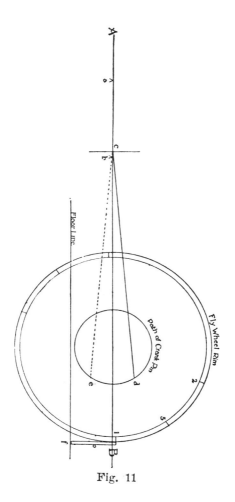

Fig. 11

fixed point on the floor, as standing a
two foot rule on end on the floor, as
shown at 1 f in the figure, and marking
over the other end of the rule where it
comes in contact with the wheel rim ;
this point is shown at 1 in the figure,
also mark the exact point on the
floor measured from. Now turn the
engine over past the center until the
lines on the cross-head and the guide
again make one continuous line (point
c in the figure), when the crank-pin
will be in a position about like e in
the figure, and make a second mark on
the fly-wheel rim, represented by point
2 in the figure. Now with a pair of
dividers, or in any other convenient
way, locate a point on the wheel rim
exactly central between points 1 and 2
and make a prick-punch mark, this
point we will call 3, see figure. By
bringing point 3 squarely to the end of
our two foot rule when the latter is
stood on end as before, the engine will
be most accurately centered. The opposite center is located by going
through the same operation with the
engine at the other end of the stroke.

If the fly-wheel runs conveniently

near to the bed plate or any permanent part of the engine frame, a reference point may be *permanently* located thereon, and used whenever desirable by making a permanent tram of a piece of stiff steel wire, thus making it the work of but a moment or two to locate the exact dead centers, after once locating and marking them.

CHAPTER VII.—SETTING THE ECCENTRIC.

A study of a few of the movements of the slide-valve as compared with the piston movement will clear up whatever apparent mystery there may be about the position of the eccentric.

In Chapter I. it was stated that if a slide-valve has neither steam-lap nor lead, the eccentric must be set at an angle 90 degrees in advance of the crank.

The travel of a slide-valve without lap or lead is equal to twice the width of the steam port; if the valve *has* steam lap, its travel must then equal twice the width of the steam port *plus* twice the steam lap on one end. Knowing these facts it is easily apparent that when the piston is at one end of its stroke the valve must—in the case of no lap nor lead—be at mid travel, or more plainly, it must have been carried forward just half way in the direction of the next piston movement, so that it may be ready to admit steam to the cylinder at the proper time, therefore

it is obvious that the eccentric must
also be at about half of its stroke, or

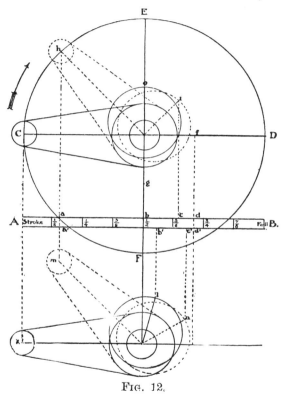

FIG. 12.

one-quarter of a revolution in advance
of the crank-pin.

Referring to Fig. 12, which shows the
relative positions of the crank-pin and

the eccentric, it will be seen that when the crank-pin is moving from C toward E, the eccentric is moving from e toward f, and when the crank-pin has arrived at E the eccentric will have reached point f, which is its extreme travel in that direction—i. e., toward the right in the figure—and when it is in this position the steam port has full opening for admission. As the crank-pin continues on its revolution the steam port is gradually closing until the eccentric has arrived at g, at which moment the crank-pin is on the other dead center, the steam port closed and the valve ready to open for admission into the other end of the cylinder for the return stroke. As steam lap is added to the valve for the purpose of working steam expansively, the eccentric must be advanced to an angle greater than 90 degrees ahead of the crank, to bring the valve into position for opening at the proper time, and as "lead" is given to the valve, this advance must be still further increased.

It is a well-known fact that the reciprocating motion derived from a crank, or other equivalent rotary motion, is intermittent; for instance, an engine piston starting from the end of its stroke accelerates in speed up to mid stroke, beyond which point its motion is retarded until it comes to a state of rest on the other center, its fastest travel being when the crank is about perpendicular to the center line of the engine. An eccentric is simply a crank with an abnormally large crank-pin, and the characteristics of the motion imparted by it are identical with that derived from a crank. The particular point which we desire to bring out being that the eccentric also will transmit its fastest motion to the slide-valve, or to the wrist-plate of a Corliss engine, as the case may be when it, the eccentric, is at a right angle to the center line of the engine, regardless of its position relative to the crank. The foregoing facts apply equally to the Corliss valve motion as to the slide-valve.

It is essential that the steam valves should move very rapidly in opening so as to give *full port opening* early in the piston stroke, therefore the fastest motion of the wrist-plate is desired when the piston is just beginning its stroke, and to attain this the eccentric must be as nearly perpendicular to the crank as is possible.

Referring again to Fig. 12, in which the parallel lines A B represent the stroke of the piston, therefore twice the length of the crank, it will be seen that with the eccentric set at 90 degrees ahead of the crank, the crank-pin having moved through one-eighth of a revolution as indicated by h, and the eccentric to the position i, the piston, in moving through a trifle more than one-eighth of its stroke has moved the eccentric, consequently the wrist-plate, through about two-thirds of its effective travel, as regards steam valve opening, as from b to c on line A B, while for the next equal movement of the crank-pin, i. e., from h to e the wrist-plate has moved only about half as far as it did for the first eighth of a revolution of the crank-pin, its total movement toward opening the steam valve being b c d. It is apparent that if the steam valve is not released for cut-off before the eccentric reaches the extreme of its travel, point f in the figure, it will not be released in that revolution, because the motion of the eccentric after passing f is in the opposite direction, therefore the crab claw will be *receding from* the knock-off cam.

Referring to the lower half of the figure, we find the same crank and the same eccentric, but sufficient lap has been given to the steam valves to require the advancing of the eccentric 15 degrees further ahead than before, or to a position 105 degrees ahead of the crank-pin, see l, in the figure.   It will be seen here that with the crank-pin moved forward one-eighth of a revolution as before—see m—the eccentric has moved from b^1 to c^1 which is considerably less than from b to c as when in its first position, and that the remaining portion of its travel during which the steam valve may be released is smaller still as shown at $c^1 d^1$; the total movement during which cut-off may take place being proportional to $b^1 c^1 d^1$, which is considerably shorter than with the first setting.

The effect of advancing the eccentric beyond 90 degrees will be that it will require a smaller load to prevent cut-off taking place, or to "make the en-

gine take steam full stroke'' than is
required to do so when set at 90 degrees.

Having placed our engine on the
dead center, say with the piston in the
head end of the cylinder, and found by
referring to the ''Table of Laps and
Lead'' that a 20 inch engine requires
1-32 inch lead, we are ready to go ahead.

ENGINE HOOKED IN.

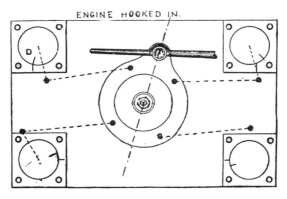

Fig. 13.

Let an assistant slowly turn the eccentric
around the shaft in the direction the
engine is to run, the reach-rod, or hook-
rod, as it is also called, being engaged
on the wrist plate, until the lap of the
steam valve on the head end is taken
up which will be indicated by the marks

on the valve and chamber being line and line, as at D, Fig. 13. Now take up a pair of dividers the 1-32 inch of required lead, and placing one leg in one of the lines, have the eccentric advanced until the line on the valve is 1-32 inch nearer the crank than that on the chamber, provided the valve rotates toward the center of the cylinder in opening, as is the case in Fig 13. Of course, if the valve opens outward the line on the valve must be on *the other side* of the one on the chamber the distance required. Fasten the eccentric, being careful that it has not been moved along the shaft, and then turn the engine on to the crank end dead center, and see if the crank end steam valve has the required opening, as it undoubtedly will if due care has been given to all the preliminary adjustments. While this is being done it will be well to see that the exhaust valves are properly lapped when engine is on the center, this lap should be the same for each end.

Having found everything to be correctly adjusted the back bonnets may now be put on the valve chambers, a careful examination made of all parts of the valve gear to see that there is no bind or interference. This being done and the eccentric securely fastened and its position on the shaft *lightly* marked, we are ready to adjust the governor.

CHAPTER VIII.—ADJUSTING THE GOVER-NOR.

Have the engine unhooked, then block up the governor three-eigths of an inch and place the wrist-plate at very nearly its extreme throw toward the frame end, thus pulling the head end steam valve almost wide open. Now adjust the cam rod which connects with the cam-collar on the head-end to such a length as will cause head-end steam valve to be unhooked when the wrist-plate is moved exactly on to its extreme throw, as will be indicated by the marks on the wrist-plate hub and stud. Having fastened the cam-rod to the head-end, put an extra quarter inch piece of blocking under the governor—thus raising it a total distance of five-eighths of an inch—and make the crank end cam-rod of such length as will cause the steam valve in this end to be released when the wrist plate is moved over to its extreme travel toward the head-end.

The reason for raising the governor

higher when adjusting the crank-end cut-off, is to make correction for the error due to the angularity of the connecting-rod. This will be explained later on in the present chapter.

The governor should now be blocked up to its extreme height, and when in this position the valves should not hook up. This will prevent the engine from "running away" should the main belt or line-shaft break, thus relieving the engine of its load.

Several of the Corliss engine governors have a collar fitted to the upright governor spindle, several inches above the counter-weight, and held in position by a set-screw ; this collar should in all cases be secured high enough up to allow of the governor being raised high enough to prevent the steam valves hooking on, but not so high as to allow the governor to be pushed far enough up as to raise the guide blocks out of the slots in the column.

In addition to the knock-off cams on the cam-collars, there will be found adjustable buttons. When the governor is resting on the safety-stop—which consists of a removable pin in the side of the governor column, or a notched collar loosely fitted around the column near its top—these safety stop buttons should be adjusted so that they will just clear the hook, thus preventing the steam valves from hooking up should the governor drop to its lowest point, through the breaking or running off of the governor belt, when the engine is running. Of course this safety collar *must* be turned, or the pin removed, as the case may be, as soon as the engine is up to speed, for if not, and the governor belt should run off or break, serious results would undoubtedly follow, because the engine would take steam full stroke as soon as the governor ran slow enough to prevent cut-off taking place.

Reference has been made to the disturbance of the cut-off, due to the angularity of the connecting-rod ; this effect is explained as follows : In Fig. 14 let A B represent the travel of the crosshead pin—consequently the piston travel—and the circle C E D F the path of the crank-pin. Assuming the crank to be on its inboard dead centre—or in the position O C—the distance A C will

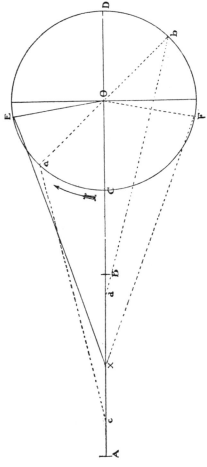

Fig. 14.

obviously be the length of the connecting-rod. If we now assume the crosshead to be in the centre of its travel, as at X, the crank-pin will have moved in the direction indicated by the arrow, to point E, which is plainly *less* than 90 degrees, and when the crosshead has travelled the same distance *on the return stroke*, the crank-pin will have travelled the space D F, which is *greater* than 90 degrees, consequently the piston travels further during the first quarter of the crank's revolution, starting from C, than it does during the second quarter ; also a *shorter* distance during the third quarter than it does during the last.

Suppose the engine to be turning over very slowly, and the governor blocked up to cut off the steam when the crankpin has made one eighth revolution, as at a on the outward stroke and b on the return stroke, it is evident, with no correction of the governor, that when cutoff takes place, the piston will have travelled the distance A c on the outward stroke, and the distance d B on the return stroke, therefore it is apparent that the point of cut-off in the

crank end is much shorter than in the
head end, as will be seen by comparing
the space d B with A c.

By putting the extra thickness of
blocking under the governor when the
crank end cam-rod is being adjusted,
the knock-off cam is moved relatively
further away from the circular limb of
the crank end crab-claw, thus allowing
this crab-claw to be moved further
toward the head end before being unhooked than would have been the case
had not the correction been made.

When the cut-offs are equalized, the
steam valves will not be released in the
same revolution when starting up the
engine; the head end valve will begin
to be released probably two or more
revolutions before the crank end valve
is unhooked, before the engine has got
up to speed. The object of equalizing
the point of cut-off in the two ends of
the cylinder is to assist in delivering as
nearly as possible a uniform rotative
effect to the belt wheel, which will assist
in perfect regulation. It must be understood that equalizing the point of
cut-off alone does not by any means
signify that each end of the cylinder

will be doing an equal share of work,
for the piston rod diminishes the effective area of the piston in the crank end
of the cylinder, and when "balancing
the load," this must be accounted for
There are still other factors which enter
into the question of stable regulation,
such as steam pressure, speed, weight
of reciprocating parts, and flywheel,
also the manner of connecting an engine
with its work.

The instructions given in this volume
if carefully followed, will result in
as nearly perfect adjustment as it is
possible to attain under ordinary conditions. Different conditions of load,
class of work, etc., will have their modi-
fying effect, and the only way to determine what the required refinements of
adjustment are to be in each case, is to
apply the indicator and abide by its
dictation.

After getting the engine to work with
its full load, should it be found, by applying the indicator, that the head end
still has the longest cut-off, the cam rod
to the crank end steam valve should
bestill further shortened if the engine
has the crab-claw gear ; should it be

equipped with the oval arm gear opening the steam valves outward, the cam rod must be lengthened—letting the head end cam rod remain as adjusted before starting up, for all error caused by the angularity of the rod must be compensated for at the crank end. When the engine is shut down again after making this second connection, it may be necessary to readjust the safety stop cam on this end, for as the engine is slowing down and the governor descending, this cam may come into play too early, thus preventing the crank end valve hooking on when the governor gets down onto the safety collar.

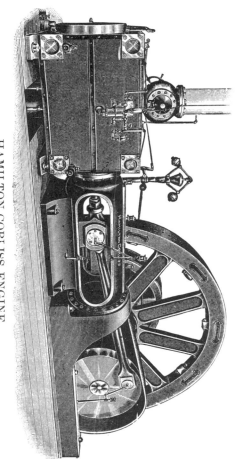

HAMILTON-CORLISS ENGINE.

CHAPTER IX.   INDICATOR DIAGRAMS.

In the preceeding chapter we referred to applying the indicator to determine the final adjustment of the valves. Let us first study the essential features on an indicator diagrams, by referring to Fig. 15, which has been

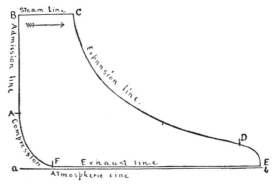

Fig. 15.

drawn by hand for illustration only, it being too near perfect for actual practice. The names of the different lines are plainly marked in the figure. The sequence of events in the cylinder for one revolution is as follows:—
The engine being on a dead center the steam enters the admission valve at that end at A in the figure, and raises

that pressure in the cylinder to B when the piston starts on its forward stroke—in the direction of the arrow. Steam "follows" the piston, at full pressure, to C, the "point of cut-off," at which time the steam valve is released by the action of the governor, thus cutting off the supply of steam, and the balance of the stroke is made by the expansive force of the steam, as shown by the "expansion curve." At D, a trifle before the piston reaches the end of its stroke the exhaust valve is opened and the expanded steam is expelled into the atmosphere. This early release greatly assists in reducing the back-pressure on the return stroke.

At E the piston starts on its return stroke—impelled from the other end in the manner just described—against the "back pressure," which is the pressure in the exhaust pipe, up to point F where the exhaust valve closes, and the piston in completing its return stroke, compresses the confined steam thus bringing the reciprocating parts up gradually for their reversal of motion, to A where the steam valve is again opened for admission. This cycle repeats itself in each end of the cylinder alternately.

The compression of the exhaust steam remaining in the cylinder at the closing of the exhaust valve, at F, not only serves to "cushion" the reciprocating parts, but it also diminishes the quantity of steam that would otherwise be required to fill the clearance volume at each stroke, thus reducing the quantity of steam required for the engine per horse-power per hour.

The line a l is the "atmospheric line," and denotes the pressure of the atmosphere at the time the card was taken, and is always equivalent to 0 pounds gauge pressure or "14.7 pounds absolute" i. e. 14.7 pounds above perfect vacuum.   It is drawn by the indicator immediately after taking a card and while the spring is still hot, with steam shut off from the instrument.

The pressures indicated by the different lines of the diagram are measured from the atmospheric line with the scale of the spring used in taking the diagram.   Thus if a 50 spring was used, and the steam line near B stood 90 points, on the 50 scale, above the atmospheric line, the "initial pressure" would be 90 pounds.

The terminal pressure, which to a great extent indicates the degree of economical performance, is measured from the point of release, D, to the atmospheric line.

The proportion of the whole length of the diagram held by the distance C from the admission line—or a line erected perpendicular to the atmospheric line and forming a part of the admission line,—represents the proportion of the engine stroke completed when cut off takes place.

In practice you will rarely get such sharply defined points as shown in the

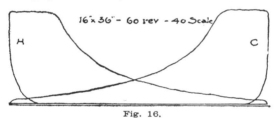

16"x 36"- 60 rev - 40 Scale

H

C

Fig. 16.

figure, unless it be at very slow speeds, they being slightly obscured by the rounding of corners, due to the comparatively gradual action of the steam in changing from one operation to another.   This gradual merging of one line into another is illustrated in the reproductions of actual diagrams shown in this chapter.

When perfecting the valve adjustment after the usual full load has been put on, the cards should be made to

approach the ideal diagram as closely
as is consistant with other conditions.

Figure 16 was taken from a 16x36 in.
Corliss engine making 60 revolutions
a minute, and is a splendid card. The
initial pressure is 63 pounds, (scale 40)
the terminal 4 pounds, and the back
pressure 1 pound, all gauge pressures.
This diagram is all that could be desired, and gives every indication of
economical performance.

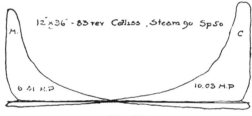

Fig. 17.

Figure 17 was taken from a new engine, Corliss type, 12x36 in.—83 revolutions. Steam pressure 90 pounds,
spring 50. The load was not all on,
as some of the machinery was not
ready to run, and the load is far too
small for the best results.

The valves had been set according
to the method described in the previous chapters, and shows what may
be accomplished by careful work,
when "setting to marks."   A few

slight corrections are necessary, no-
tably the rather late admission in the
head end, as shown by the inclination
of the admission line toward the center of the diagram.   No adjustments
were made at this time, it being decided to wait for the full load before
making any corrections.

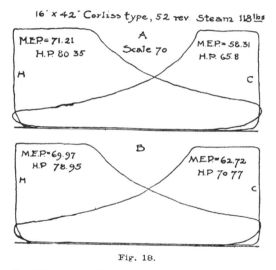

Fig. 18.

In Figure 18, card A was taken from
a new Corliss type "straight line" air
compressor—air and steam cylinders
tandem,— the valves on this engine
also were set to marks.   It illustrates
the effect of the angularity of the connecting rod upon the point of cut off,

as described in chapter viii, the head
end indicating 14:55 horse power more
than the crank end.   Card B was
taken a few minutes later after lengthening the governor rod to the crank
end knock off cam,—the "oval arm
gear" being used on this engine—with
the result that the difference in load
between the ends of the cylinder was
reduced to 8.18 horse power.   The
load was balanced within a fraction of
a horse power when the total load was
about 140 horse power before leaving
this engine, but the final diagrams
have been mislaid.

A little more compression and an
earlier release would have been beneficial to this engine, in fact a slight advancement of the eccentric thus making each function of the valves earlier,
would not be much amiss.

CHAPTER X.   A FEW POINTERS.

When starting a new engine for the first time the greatest care should be exercised. Get the cylinder and valves thoroughly warmed up before the engine is started and when you do start do not hook on the valve gear but run several revolutions moving the valves by hand, or as it is usually called, "with the bar," observing the action of the valve gear and other small parts while doing so. When you are positive that there is no bind or interference anywhere, hook on the valve gear and allow the engine to run slowly for several minutes, then get it gradually up to speed.

Do not try to economize in the use of oil for the first few days; use plenty of *good* cylinder oil in the cylinder. The surfaces of the cylinder and valves will be improved by the application of Dixon's flaked graphite prepared for this purpose, which can be mixed with cylinder oil and injected with a hand pump, or fed clear in a cup especially designed for it. This graphite is an excellent antidote for hot bearings, besides being exceeding-

ly useful in many other ways, and
should be included in the list of sup-
plies for the engine room.

During the trial run demonstrate
the efficiency of all safety appliances
and *know postively* that they are adjusted so as to perform that which
they were designed for; in fact never
assume anything to be all right when
dealing with any of the various forces
existing in a steam plant, but *know* by
actual investigation.

Do not try to "key up" the brasses,
or adjust any of the bearings, to the
utmost nicety for a few days so long
as they do not pound; it is better to
run a trifle slack until they have attained a "surface"; in other words,
better a little noise than a hot bearing.

After a few hours' run examine all
fastenings to determine if any of them
are inclined to work loose, and after a
few days of actual running with the
load on, examine the anchor bolts to
see if any of them have become slack;
take off the cylinder head and examine the follower bolts and piston rod
nut, to make sure they are going to
stay where the belong. This examination of the cylinder should be
made three or more times a year, and

the piston kept in the center of the cylinder. Also keep the cross head so adjusted that the piston rod shall always be concentric with the center line of the engine.

In keying up the connecting rod brasses it should be remembered that the equality of the clearance in the ends of the cylinder is gradually destroyed, and if no correction is made the piston will in time be brought up against one of the cylinder heads, according to the kind of rod and method of adjusting. In engines whose connecting rod ends are fitted with the usual straps and keys the repeated driving of the keys shortens the effective length of the connecting rod, thus diminishing the clearance in the crank end—or "back end"—of the cylinder. This is corrected by interposing sheet steel liners or "shims" between the stub ends of the rod and the inside brasses thus maintaining a nearly constant length of rod. With "solid end" rods keying up lengthens the rod thus diminishing the clearance in the head end of the cylinder, therefore the shims in this case should be put between the extreme ends of the rod eyes and the outside brasses. Ordinarily it will take a very long time to sensibly alter the clearance, but it should be looked into occasionally by referring to the "striking points" laid off on the guides.

After the normal load is all on, the engine settled right down to business, and the valve adjustment corrected with the indicator to conform to the conditions under which the engine is to run, mark the eccentric's position permanently upon the shaft, also

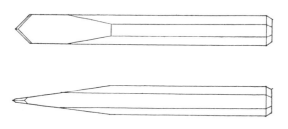

Fig. 19.

mark the position of the eccentric rod and hook rod ends, wherever there is a screwed joint, to conform to the tram gauge which was laid off on the frame when the reciprocating parts were being set up (Chapter VII. part I.).

For marking eccentrics—or any similar part of any machine—we have found the tool illustrated in Fig. 19, very handy. It is made like a cold chisel with the exception that it has two cutting edges at a right angle, the apex of the angle being directly in the longitudinal center of the chisel. On using, its point is placed in the angle formed by the eccentric and the shaft, when a light blow with a hammer marks both shaft and eccentric at the same time, with the marks in exact line.

Occasionally a Corliss engine, especially those built ten or a dozen years ago with the old style slow speeded governor, will have an unaccountable fit of racing, when everything connected with the regulating mechanism is apparently in the best of condition. Well do we remember an instance of this kind in our early experience which completely baffled us yet it was almost immediately remedied after we had called in a brother engineer from a neighboring plant. The collar which takes the weight of the vertical spindle together with the balls, is located some six or eight inches down inside the governor column, just below the bottom of the guide slot. When our neighbor entered the engine room and observed the antics of the governor, he asked for a squirt can full of kerosene, and he pro-

ceeded to flush this collar with it.   In
a few revolutions the racing had stopped and after cooling down this bearing, which had got quite warm, no
further trouble was experienced. The
fact is these collars are rather difficult
to properly lubricate, and in the case
referred to whatever oil had found its
way there had gummed, thus the friction. The best oil which we have found
to use upon any governor of this type
is called "high viscosity spindle oil":
it is a very fluid, light colored, neutral
oil, and, as its name implies, has good
lubricating   properties,   and   being
strictly mineral, it will not gum.

One of the common faults of the
crab claw gear which will cause racing, and send the un-initiated on a
wild goose chase, is occasioned by the
unavoidable wearing of the "steels."
When the steel hook contacts are
new the circular limb of the crab claw
is concentric with the center of the
hook block stud.   The wearing of the
contact edges of the block and steel
and the consequent grinding up the
same, changes the relative position of
the hook, block, and knock-off cam to
such an extent that the cam cannot
release the hook if the load is such
as will require a late cut off, thus

causing the engine to take steam full
stroke for a revolution or two and
making it race. Referring to Fig. 20,
the arc c shows the changed relation
of the circular limb of the hook to the
other parts, after the steels have been
shortened by wear and grinding. This
new position as shown by arc c shows
why this trouble is confined more to a
late cut off than an early one.

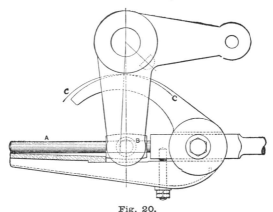

Fig. 20.

The way to determine the extent of
wear, consequently the required
length of new steel (the section-lined
portion in the figure represents the
steel) is as follows:— Place the point
of an hermaphrodite caliper in the
center of the hook block stud and with
the other end of the caliper follow the

outline of the circular limb of the
claw. If the caliper does not follow
the curvature of the claw, but runs off,
following a path similar to c in the figure, slide the block along the spindle
(using a hand clamp to lift it against
the action of the dash pot and hold it
steady for measurement) until the caliper will follow the arc of the circular
limb; then the distance between the
hook block and the steel on the claw
shows how much longer the new steel
should be. It will do no harm to make
the new steels, say, a thirty-second of
an inch longer than this, to allow for
wear. The steels should be hardened.

Another fault of this gear, which
makes itself apparent through negli-
gence, is the tendency of the block B
to stick fast to the spindle A, due to
lack of lubrication or the use of an
inferior quality of engine oil. When
this sticking occurs it is evident that
the dash pot cannot close the valve,
consequently the engine must take
steam full stroke, and the governor is
powerless to prevent it, thus it is apparent that a catastrophy is immin-
ent if it is not discovered in time. In
fact, it is positively known that fly
wheels have burst, engines been

wrecked, buildings been demolished,
and lives destroyed from this cause.

In any engine room which houses
an engine whose cylinder is twelve
inches or over in diameter, there
should be eyebolts permanently placed
in the ceiling, one over the center line
of the engine say four inches back
from the cylinder head, so that the
head may be hoisted or lowered clear
of the studs whenever it is necessary
to examine the cylinder, another one
over the middle of the connecting rod,
and others over each main bearing.
Their cost is trifling and they save
time and labor.

As to tools for emergencies:—we
have found the following list very ap-
propriate. It should be selected according to the size and weight of the parts
to be handled but ordinarily the sizes
named will be about right:— One
chain hoist of 1500 pounds capacity, a
tackle with one single and one double
shive blocks with five-eighth inch (diameter) rope, two 12 inch or 14 inch
screw jacks, a good hickory lever 8 or
10 feet long and 4 inch to 6 inch at the
butt, a crow bar, small pinch bar, and
an assortment of rope slings and
blocking. We find that provision in this
line is woefully lacking in the major-

ity of engine rooms which we have had the opportunity of visiting. These things may seldom be needed, but in one emergency job—and accidents *do* sometimes happen—they will usually more than pay their cost through the amount of time and labor saved.

CHAPTER XI. THE DOUBLE PORTED VALVE, AND THE "LONG RANGE CUT-OFF."

The two most important improvements in the Corliss valve gear are the double ported valve and the adoption of separate eccentrics for the steam and exhaust valves.

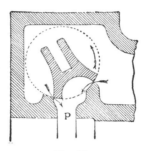

Fig. 21.

Fig. 21, is a sectional view of the "double ported" steam valve and valve chamber and shows the relative position of the working edges of the valve and ports. The valve moves in the direction of the short arrow in opening, and steam enters the port P as indicated by the crooked arrows. The steam is admitted and remains at nearly full boiler pressure up to point of cut-off, the latter being very pronounced on the indicator diagram, when this style valve is used.

While the idea of using two eccentrics and two wrist plates, can hardly be considered as a recent improvement, they were not generally adopted until a few years ago.

As long ago as 1877 several engineers realized the benefits to be derived from separating the driving mechanism of the steam and exhaust valves, and begun agitating the matter but for some reason builders generally refused to adopt the idea. In 1886 a few builders began to equip the low pressure cylinders of compound engines with separate eccentrics for the steam and exhaust valves, but still using a single eccentric for the high pressure side. The fallacy of this arrangement soon became apparent, as when a good load was put upon the engine the low pressure cylinder would empty the receiver, owing to the contracted range of valve movement on the high pressure cylinder not furnishing a sufficient quantity of steam, therefore both cylinders were finally equipped with two eccentrics.

At the present time any of the leading builders will equip engines with two eccentrics when specified, and

several of them make a specialty of
regularly furnishing engines, either
simple or compound, with the double
eccentric valve gear. Fig. 22 illus-

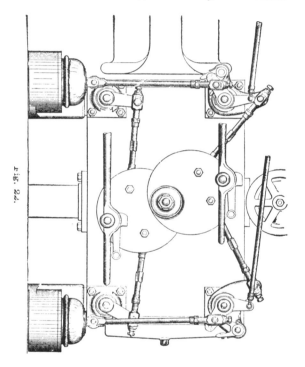

Fig. 22.

tes the valve gear of a well known
make of Corliss simple engine with
two eccentrics, for "long range cut-
off."

A Corliss engine with a single eccentric, having the valves adjusted and eccentric set so as to give the greatest range of cut off—i e, with the least possible angular advance—if put to work under a heavy load may be handicapped by its inability to exhaust the steam early enough to bring the exhaust down to atmosphere pressure at the beginning of the return stroke, or as the saying is, to "get rid of the steam," and if the eccentric be advanced to secure early release the range of cut off under control of the governor is so reduced that the steam valves may not be released for one or two strokes, thus augmenting the trouble which it was desired to remedy.

With separate eccentrics for the steam and exhaust valves, the exhaust eccentric may be given a good advance, thus securing an early release and sufficient compression to fill the clearance space, and warm the cylinder walls up to a temperature approaching that of the entering steam, while the steam eccentric may be so set as to have the laps of the steam valves taken up when this eccentric is set with *negative* angular advance, thus giving a great range of cut off, and the greatest range of power under control of the governor.

The setting of the steam eccentric varies from 9 degrees *negative* angular advance to 6 degrees positive angular advance or from 81 degrees to 96 degrees in advance of the crank, according to the requirements of the case. The 9 degree negative advance position provides for about seven tenths cut off.

CHAPTER XII. TABLES AND MEMOR- ANDA.

NOTES ON STEAM AND FUEL CONSUMP- TION.

A good many automatic non-condensing engines require from three to four pounds of coal per horse-power per hour, according to the quality of the coal and the efficiency of the boiler. An automatic condensing engine requires from two and one-quarter to three and one-half pounds of coal per horse-power per hour. A steam-jacketed compound condensing engine of the most improved construction may, with good management, reduce the consumption of coal as low as one and three-quarters to two pounds of coal per horse-power per hour.

The average amount of feed water required for a good, economical engine, is about 26 pounds per indicated horse-power per hour; engines of high economy—compound and triple expansion—will use less than this amount. A high piston speed, together with a high rotative speed, is very desirable, as a great power may thus be obtained from the moderate sized engines, and the evil of cylinder condensation corrected to a great extent, but these are somewhat limited by practical considerations.

A good condenser increases the economical efficiency of an engine from twenty-five to forty per cent., and the amount of injection water required for condensing may be roughly taken at about twenty-five times the quantity fed to the boilers.

In estimating for a consumption of fourteen pounds of coal per square foot of grate per hour, about eight pounds of water may be taken as the rate of evaporation per pound of coal, which can be done with good natural draft. With forced draft and twenty-eight pounds of coal per square foot of grate, the evaporation is only about six pounds of water to one of coal.

Each pound of coal per hour is:—
1.5   net tons per year of 300, 10h. days
1.34 gross " " " " " " "
3.6   net " " " " 24h "
3.21 gross " " " " " " "

With eight pounds of water evaporated per pound of coal, each pound of steam (water) per horse-power takes:
.1875 net tons per year of 300, 10h. d'ys
.1675 gross " " " " " " "
.45   net " " " " " 24h "
.4   gross " " " " " " "

HORSE POWER OF AN ENGINE.

Formula:—

$$H.\ P. = \frac{PLAN}{33000}$$

P=mean effective pressure on the piston.

L=the length of the stroke of engine in feet.

A=area of the piston . in square inches.

N=number of strokes of piston in a minute.

33000=foot pounds of work equal to one horse-power.

PROPERTIES OF SATURATED STEAM.

Ice is liquefied and becomes water at 32 degrees F. Above this point water increases in temperature up to the steaming point, nearly at the rate of 1 degree for each unit of heat added per pound of water. The steaming point (212 degrees at atmospheric pressure) rises as the superimposed pressure increases.

For each unit of heat added above the steaming point, a portion of the water is converted into steam, having the same temperature and the same pressure as that at which it is evaporated. The heat so absorbed is called "Latent Heat." The amount of heat rendered latent by each pound of water in becoming steam varies at different pressures, decreasing as the pressure increases. The latent heat, added to the sensible heat (or thermometric temperature), constitutes the "Total Heat." The "total heat" being greater as the pressure increases, it will take more heat, and, consequently, more fuel, to make a pound of steam the higher the pressure.

The table on page 143 gives the properties of steam at different pressures —from 1 lb. to 400 lbs. "total pressure," i. e., above vacuum.

The gauge pressure is about 15 pounds less than the total pressure, so that in using this table, 15 must be added to the pressure as given by the steam gauge.

The column of Temperatures gives the thermometric temperature of steam and boiling point at each pressure.

The "factor of equivalent evaporation" shows the proportionate cost, in heat or fuel, of producing steam at the given pressure, as compared with atmospheric pressure. To ascertain the equivalent evaporation at any press-

ure, multiply the given evaporation by
the factor of its pressure, and divide
the quotient by the factor of the desired pressure.

Each degree of difference in temperature of feed water, makes a difference
of .00104 in the amount of evaporation. Hence, to ascertain the equivalent evaporation from any other temperature of feed than 212 degrees, add
to the factor given as many times
.00104 as the temperature of feed water is degrees below 212 degrees.

For other pressures than those
given in the table, it will be practically correct to take the proportion of the
difference between the nearest pressures given in the table.

MEMORANDA ON WATER.

1 cubic foot of fresh water at maxi-
mum density, 39.2 degrees F. weighs
62.48 lbs.

I cubic inch of fresh water at maxi-
mum density, 39.2 degrees F. weighs
.03617 lbs.

1 cubic foot of fresh water at boiling
point, 212 degrees F. weighs 59.76
lbs.

1 cubic foot of fresh water at standard
temperature, 62 degrees F. weighs
62.355 lbs.

35.84 cubic feet of fresh water weighs 2240 lbs.

1 cubic foot of fresh water contains 7.48 U. S. Gals.

1 U. S. Gallon of fresh water weighs 8.35 lbs.

1 U. S. Gallon of fresh water contains 231 cu. in.

1 Pound of fresh water at 62 degrees F. contains 27.64 in.

PRESSURE OF A COLUMN OF WATER.

A column of water one foot high exerts a pressure of .434 pounds per square inch, therefore to ascertain the pressure per square inch upon the base of a column of water, multiply its height in feet by .434 pounds.

H. P. REQUIRED TO ELEVATE WATER.

To determine the horse power necessary to elevate water to a given height, multiply the number of gallons per minute by 8.35, the weight of one gallon; multiply this product by the total number of feet the water is raised, and the last product will be the foot-pounds of work done in one minute. Divide this quantity by 33,-000; the quotient will be the net horse power, to which add twenty-five per centum for friction, slip, etc.

CONVENIENT APPROXIMATE MULTIPLI- ERS.

Square inches x .007==square feet.

Square feet x .111==square yards.

Cubic inches x .00058==cubic feet.

Cubic feet x .03704==cubic yards.

Cubic inches x .004329==U. S. gallons.

Cubic feet x 7.48==U. S. gallons.

Cubic feet x 62.355==pounds.

U. S. gallons x 231.==cubic inches.

U. S. gallons x .13368==cubic feet.

Diameter of a circle x 3.1416==circum- ference.

Diameter of a circle x .8862==side of equal square.

Circumference of a circle x .31831== diameter.

Square of diameter of circle x .7854== area.

AREAS OF CIRCLES.

Areas of Circles having Diameters varying from 1 Inch to 100 Inches.

Diam. in Inches	Area in Square Inches.	Diam. in Inches.	Area in Square Inches.	Diam. in Inches.	Area in Square Inches.
1	0 7854	3 3/16	7 9798	5 3/4	25 967
1 1/16	0 8866	3 1/4	8 2957	5 7/8	27 108
1 1/8	0 9940	3 5/16	8 6180	6	28 274
1 3/16	1 1075	3 3/8	8 9462	6 1/8	29.464
1 1/4	1 2271	3 7/16	9 2807	6 1/4	30 679
1 5/16	1 3530	3 1/2	9.6211	6 3/8	31 919
1 3/8	1 4848	3 9/16	9.9680	6 1/2	33.183
1 7/16	1 6229	3 5/8	10 320	6 5/8	34 471
1 1/2	1 7671	3 11/16	10.679	6 3/4	35 784
1 9/16	1 9175	3 3/4	11.044	6 7/8	37 122
1 5/8	2.0739	3 13/16	11.416	7	38 484
1 11/16	2 2365	3 7/8	11.793	7 1/8	39.871
1 3/4	2.4052	3 15/16	12.177	7 1/4	41 282
1 13/16	2.5800	4	12.566	7 3/8	42.718
1 7/8	2.7611	4 1/16	12.962	7 1/2	44 178
1 15/16	2.9483	4 1/8	13.364	7 5/8	45.663
2	3 1416	4 3/16	13.772	7 3/4	47.173
2 1/16	3 3380	4 1/4	14.186	7 7/8	48.707
2 1/8	3 5465	4 5/16	14.606	8	50.265
2 3/16	3.7584	4 3/8	15.033	8 1/8	51.848
2 1/4	3 9760	4 7/16	15.465	8 1/4	53.456
2 5/16	4.2000	4 1/2	15.904	8 3/8	55.088
2 3/8	4.4302	4 9/16	16.349	8 1/2	56.745
2 7/16	4 6664	4 5/8	16.800	8 5/8	58 426
2 1/2	4 9087	4 11/16	17.257	8 3/4	60 132
2 9/16	5 1573	4 3/4	17 720	8 7/8	61.862
2 5/8	5.4119	4 13/16	18.190	9	63.617
2 11/16	5.6723	4 7/8	18.665	9 1/8	65.396
2 3/4	5.9395	4 15/16	19.147	9 1/4	67.200
2 13/16	6.2126	5	19.635	9 3/8	69.029
2 7/8	6.4918	5 1/16	20 629	9 1/2	70.882
2 15/16	6.7772	5 1/4	21.647	9 5/8	72.759
3	7.0686	5 3/8	22.690	9 3/4	74.662
3 1/16	7.3662	5 1/2	23.758	9 7/8	76.588
3 1/8	7 6699	5 5/8	24 850	10	78.540

Diam. in Inches.	Area in Square Inches.	Diam. in Inches.	Area in Square Inches.	Diam. in Inches.	Area in Square Inches.
10⅛	80.515	14⅞	173.782	19⅝	302.489
10¼	82.516	15	176.715	19¾	306.355
10⅜	84.540	15⅛	179.672	19⅞	310.245
10½	86.590	15¼	182.654	20	314.160
10⅝	88.664	15⅜	185.661	20⅛	318.099
10¾	90.762	15½	188.692	20¼	322.063
10⅞	92.885	15⅝	191.748	20⅜	326.051
11	95.033	15¾	194.828	20½	330.064
11⅛	97.205	15⅞	197.933	20⅝	334.101
11¼	99.402	16	201.062	20¾	338.163
11⅜	101.623	16⅛	204.216	20⅞	342.250
11½	103.869	16¼	207.394	21	346.361
11⅝	106.139	16⅜	210.597	21⅛	350.497
11¾	108.434	16½	213.825	21¼	354.657
11⅞	110.753	16⅝	217.077	21⅜	358.841
12	113.097	16¾	220.353	21½	363.051
12⅛	115.466	16⅞	223.654	21⅝	367.284
12¼	117.859	17	226.980	21¾	371.543
12⅜	120.276	17⅛	230.330	21⅞	375.826
12½	122.718	17¼	233.705	22	380.133
12⅝	125.184	17⅜	237.104	22⅛	384.465
12¾	127.676	17½	240.528	22¼	388.822
12⅞	130.192	17⅝	243.977	22⅜	393.203
13	132.732	17¾	247.450	22½	397.608
13⅛	135.297	17⅞	250.947	22⅝	402.038
13¼	137.886	18	254.469	22¾	406.493
13⅜	140.500	18⅛	258.016	22⅞	410.972
13½	143.139	18¼	261.587	23	415.476
13⅝	145.802	18⅜	265.182	23⅛	420.004
13¾	148.489	18½	268.803	23¼	424.557
13⅞	151.201	18⅝	272.447	23⅜	429.135
14	153.938	18¾	276.117	23½	433.731
14⅛	156.699	18⅞	279.811	23⅝	438.363
14¼	159.485	19	283.529	23¾	443.014
14⅜	162.295	19⅛	287.272	23⅞	447.699
14½	165.130	19¼	291.039	24	452.390
14⅝	167.989	19⅜	294.831	24⅛	457.115
14¾	170.873	19½	298.648	24¼	461.864

Diam. in Inches.	Area in Square Inches.	Diam. in Inches.	Area in Square Inches.	Diam. in Inches.	Area in Square Inches.
24⅜	466.638	29⅛	666.227	37¾	1119.24
24½	471.436	29¼	671.958	38	1134.11
24⅝	476.259	29⅜	677.714	38¼	1149.08
24¾	481.106	29½	683.494	38½	1164.15
24⅞	485.978	29⅝	689.298	38¾	1179.32
25	490.875	29¾	695.128	39	1194.50
25⅛	495.796	29⅞	700.981	39¼	1209.95
25¼	500.741	30	706.860	39½	1225.42
25⅜	505.711	30¼	718.690	39¾	1240.08
25½	510.706	30½	730.618	40	1256.60
25⅝	515.725	30¾	742.644	40¼	1272.39
25¾	520.769	31	754.769	40½	1288.25
25⅞	525.837	31¼	766.992	40¾	1304.20
26	530.930	31½	779.313	41	1320.25
26⅛	536.047	31¾	791.732	41¼	1336.40
26¼	541.189	32	804.249	41½	1352.65
26⅜	546.356	32¼	816.865	41¾	1369.00
26½	551.547	32½	829.578	42	1385.44
26⅝	556.762	32¾	842.390	42¼	1401.98
26¾	562.002	33	855.30	42½	1418.62
26⅞	567.267	33¼	868.30	42¾	1435.56
27	572.556	33½	881.41	43	1452.20
27⅛	577.870	33¾	894.61	43¼	1469.13
27¼	583.208	34	907.92	43½	1486.17
27⅜	588.571	34¼	921.32	43¾	1503.30
27½	593.958	34½	934.82	44	1520.53
27⅝	599.370	34¾	948.41	44¼	1537.86
27¾	604.807	35	962.11	44½	1555.28
27⅞	610.268	35¼	975.90	44¾	1572.81
28	615.753	35½	989.80	45	1590.43
28⅛	621.263	35¾	1003.78	45¼	1608.15
28¼	626.798	36	1017.87	45½	1625.97
28⅜	632.357	36¼	1032.06	45¾	1643.89
28½	637.941	36½	1046.35	46	1661.90
28⅝	643.594	36¾	1060.73	46¼	1680.01
28¾	649.182	37	1075.21	46½	1698.23
28⅞	654.839	37¼	1089.79	46¾	1716.54
29	660.521	37½	1104.46	47	1734.94

Diam. in Inches.	Area in Square Inches.	Diam. in Inches.	Area in Square Inches.	Diam. in Inches.	Area in Square Inches.
47¼	1753 45	59	2733 97	73½	4242.92
47½	1772 05	59½	2780.51	74	4300.85
47¾	1790 76	60	2827.44	74½	4359.16
48	1809.56	60½	2874.76	75	4417.87
48¼	1828.46	61	2922.47	76	4536.47
48½	1847.45	61½	2970 57	77	4656.63
48¾	1866.55	62	3019.07	78	4778.37
49	1885 74	62½	3067.96	79	4901.68
49¼	1905.03	63	3117.25	80	5026.56
49½	1924 42	63½	3166 92	81	5153.00
49¾	1943.91	64	3216.99	82	5281.02
50	1963.50	64½	3267.46	83	5410 62
50½	2002.96	65	3318.31	84	5541.78
51	2042.82	65½	3369.56	85	5674.51
51½	2083.07	66	3421.20	86	5808.81
52	2123.72	66½	3473.23	87	5944.69
52½	2164.75	67	3525.62	88	6082.13
53	2206.18	67½	3578.47	89	6221.15
53½	2248.01	68	3631.68	90	6361.74
54	2290.22	68½	3685.29	91	6503.89
54½	2332.83	69	3739.28	92	6647.62
55	2375.83	69½	3793.67	93	6792.92
55½	2419.22	70	3848.46	94	6939.79
56	2463.01	70½	3903.63	95	7088.23
56½	2507.19	71	3959.20	96	7238.24
57	2551.76	71½	4015.16	97	7389.80
57½	2596.72	72	4071.51	98	7542.96
58	2642.08	72½	4128.25	99	7697.68
58½	2687.83	73	4185.39	100	7854.00

TABLE OF PROPERTIES OF SATURATED STEAM.

Total pressure per square inch.	Temperature in Fahrenheit degrees.	Total Heat, in heat units from water at 32° F	Latent heat, in heat units.	Density or weight of one cubic ft.	Volume of one pound of steam.	Relative volume, or cubic feet of steam from one cub. feet of water.	Factor of evaporation from water at 212°
1	102	1113.05	1042.964	.0030	330.26	20620	0.965
2	126.266	1120.45	1026.010	.0058	172.08	10720	0.972
3	141.622	1125.131	1015.254	.0085	117.52	7326	0.977
4	153.070	1128.625	1007.229	.0112	89.62	5600	0.981
5	162.330	1131.449	1000.727	.0137	72.66	4535	0.984
6	170.123	1133.825	995.249	.0163	61.21	3814	0.986
7	176.910	1135.896	990.471	.0189	52.94	3300	0.988
8	182.910	1137.726	986.245	.0214	46.69	2910	0.990
9	188.316	1139.375	982.434	.0249	41.79	2607	0.992
10	193.240	1140.877	978.958	.0264	31.84	2360	0.994
15	213.025	1146.912	964.973	.0387	25.85	1612	1.000
20	227.917	1151.454	954.415	.0511	19.72	1220.3	1.005
25	240.000	1158.139	945.825	.0634	15.99	984.8	1.008
30	250.245	1158.263	938.925	.0755	13.46	826.8	1.012
35	259.476	1160.987	932.152	.0875	11.65	713.4	1.015
40	267.120	1163.410	926.472	.0994	10.27	628.2	1.017
45	274.296	1165.600	921.324	.1111	9.18	561.8	1.017
50	280.854	1167.600	916.631	.1227	8.31	508.5	1.021
55	286.897	1169.442	912.290	.1343	7.61	464.7	1.023
60	292.520	1171.158	908.247	.1457	7.01	428.5	1.025
65	297.777	1172.762	904.462	.1569	6.49	397.7	1.027
70	302.718	1174.269	900.899	.1681	6.07	371.2	1.028
75	307.388	1175.692	897.526	.1794	5.68	348.3	1.030
80	311.812	1177.042	894.330	.1901	5.35	328.3	1.031
85	316.021	1178.326	891.286	.2010	5.05	310.5	1.033
90	320.039	1179.551	888.375	.2118	4.79	294.7	1.034
95	323.884	1180.724	885.588	.2224	4.55	280.6	1.035
100	327.571	1181.849	883.914	.2330	4.33	267.9	1.036
105	331.113	1182.929	880.342	.2434	4.14	265.5	1.037
110	334.523	1183.970	877.865	.2537	3.97	246.0	1.038
115	337.814	1184.974	875.472	.2640	3.80	236.3	1.039
120	340.995	1185.944	873.155	.2742	3.65	227.6	1.040
125	344.074	1186.883	870.911	.2842	3.51	219.7	1.041
130	347.059	1187.794	868.735	.2942	3.38	212.3	1.042
140	352.757	1189.535	864.566	.3138	3.16	199.0	1.044
150	358.161	1191.180	860.621	.3340	2.96	187.5	1.046
160	363.277	1192.741	856.874	.3520	2.79	177.3	1.047
170	368.158	1194.228	853.294	.3709	2.63	168.4	1.049
180	372.822	1195.650	849.869	.3889	2.49	160.4	1.051
190	377.291	1197.013	846.584	.4072	2.37	153.4	1.052
200	381.573	1198.319	843.432	.4249	2.26	147.1	1.053
250	401.072	1203.735	831.222	.5464	1.83	114	1.059
300	418.225	1208.737	819.610	.6486	1.54	96	1.064
350	431.956	1212.580	810.690	.7498	1.33	83	1.068
400	444.919	1217.091	800.198	.8502	1.18	73	1.073

Size of Engine.	Revs. per Minute.	Piston Speed. Ft. per Min.	H. P. Constant at 1 lb. M. E. P. for Each Revolution.	H. P. at 80 lbs. Boiler Pressure. Point of Cut Off. $\frac{1}{5}$ M. E. P. 33.58	$\frac{1}{4}$ M. E. P. 40.68	$\frac{1}{3}$ M. E. P. 50.46	H. P. at 100 lbs. Boiler Pressure. Point of Cut Off. $\frac{1}{5}$ M. E. P. 44.03	$\frac{1}{4}$ M. E. P. 52.61	$\frac{1}{3}$ M. E. P. 64.43	Fly Wheel. Diameter in Feet.	Face in Inches.	Measurements of Engines. Over All Length.	Over All Width.	Center of Shaft to Back End of Cylinder.	Center Line of Engine to Top of Finished Floor.	Diameter of Steam Pipe.	Diameter of Exhaust Pipe.
14″x36″	100	600	.0280	94	114	141	123	147	181	12	18	22′-11″	8′-6″	16′-11″	24⅞″	5″	6″
16″x36″	100	600	0363	122	148	183	160	191	234	14	19	24′-0″	8′-10″	17′-0″	26⅞″	"	"
16″x42″	100	700	0426	143	173	215	188	224	275	14	22	26′-3″	9′-1″	19′-3″	26⅞″	"	"
16″x48″	90	720	0487	147	178	221	193	231	283	14	24	28′-6″	9′-3″	21′-6″	26⅞″	"	"
18″x42″	90	630	.0541	164	198	246	214	256	314	15	26	27′-0″	9′-11″	19′-6″	28⅞″	6″	7″
18″x48″	85	680	.0618	176	214	265	231	276	339	15	29	29′-3″	10′-2″	21′-9″	28⅞″	"	"
20″x42″	85	595	0666	190	230	286	249	298	365	16	29	27′-8″	10′-7″	19′-8″	30⅞″	"	8″
20″x48″	85	680	.0761	217	263	326	285	340	417	16	33	29′-10″	10′-11″	21′-10″	30⅞″	"	"
22″x48″	85	680	0921	263	319	395	345	412	505	18	35	31′-3″	11′-8″	22′-3″	32⅞″	7″	"
24″x48″	85	680	.1096	313	379	470	410	490	600	20	25	32′-3″	11′-5″	22′-3″	33⅞″	8″	10″
24″x60″	80	800	1370	368	446	553	483	577	706	20	46	36′-9″	13′-2″	26′-9″	33⅞″	"	"
26″x48″	85	680	1287	367	445	552	482	576	705	20	28	32′-6″	11′-11″	22′-6″	34⅞″	"	"
26″x60″	80	800	1609	432	524	650	567	677	830	20	34	37′-0″	14′-1″	27′-0″	34⅞″	"	"
28″x48″	80	640	1493	401	486	603	526	678	770	20	50	32′-7″	14′-2″	22′-7″	35⅞″	9″	12″
28″x60″	75	750	1867	470	570	707	616	728	903	22	58	38′-3″	14′-10″	27′-3″	35⅞″	"	"
30″x48″	75	600	1714	432	523	649	566	676	829	22	53	33′-10″	14′-9″	22′-10″	36⅞″	10″	"
30″x60″	70	700	2143	504	610	757	660	789	967	24	62	39′-5″	15′-6″	27′-5″	36⅞″	"	"
30″x72″	60	720	2571	518	628	778	679	812	994	26	65	44′-11″	15′-9″	31′-11″	36⅞″	"	"
32″x60″	65	650	2436	532	644	799	697	833	1021	26	62	40′-7″	16′-4″	27′-7″	37⅞″	"	14″
32″x72″	60	720	2924	589	714	885	772	923	1131	27	72	45′-7″	16′-11″	32′-1″	37⅞″	"	"
34″x60″	65	650	2752	601	728	903	787	941	1153	26	72	40′-9″	17′-5″	27′-9″	38⅞″	11″	"
34″x72″	60	720	3303	666	806	1000	872	1043	1277	28	80	46′-8″	18′-1″	32′-3″	38⅞″	"	"
36″x60″	65	650	3085	673	816	1012	883	1035	1292	27	76	41′-8″	18′-9″	28′-2″	40⅞″	12″	15″
36″x72″	60	720	3702	746	904	1121	978	1164	1432	30	82	41′-6″	19′-3″	32′-6″	40⅞″	"	"
40″x60″	65	650	3809	821	1007	1249	1090	1303	1596	28	90	42′-1″	20′-1″	28′-1″	44″	"	16″
40″x72″	60	720	.4571	921	1116	1384	1207	1443	1768	30	102	47′-7″	21′-1″	32′-7″	44″	"	"

HORSE POWER AND DIMENSIONS SINGLE CORLISS NON-CONDENSING ENGINE.

CHAPTER XIII. THE REYNOLDS-CORLISS ENGINE.

This engine is built from the designs of Mr. Edwin Reynolds, in both horizontal and vertical styles, including triple and quadruple expansion engines.

Figure 23 is a view of the crank side of the Reynolds-Corliss "1890" engine, and Figure 24 illustrates the valve-gear side of the same style. The wearing surfaces are all extra large, particularly the cross-head and guides, and the engine throughout is admirably adapted for long continuous duty under the high steam pressures commonly used in electric railway and lighting stations, for which it is much used. Figure 25 illustrates a tandem compound engine of the same design.

The standard girder-frame Reynolds-Corliss engine, which is extensively used for manufacturing plants is well illustrated in Figure 26, which is a tandem compound of this pattern; the cross-compound girder-frame engine is illustrated in Figure 27.

The valve-gear was designed by Mr. Edwin Reynolds in 1876, and is one of the standard styles used at the pres-

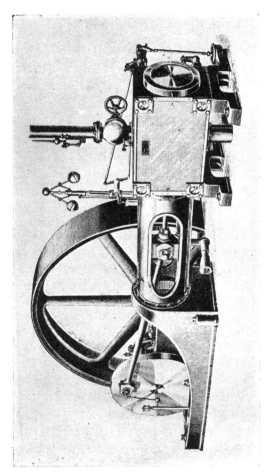

Fig. 23.

REYNOLDS-CORLISS ENGINE CRANK SIDE.

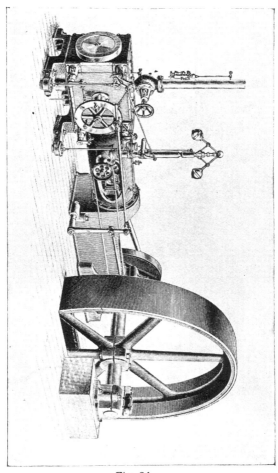

REYNOLDS-CORLISS ENGINE VALVE SIDE.

Fig. 24.

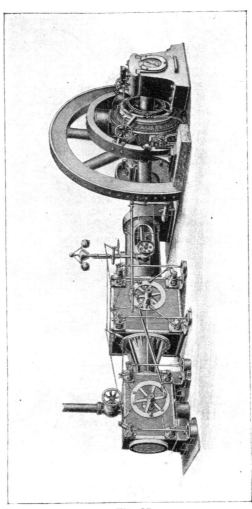

TANDEM COMPOUND REYNOLDS-CORLISS ENGINE, DIRECT COUPLED.

Fig. 25.

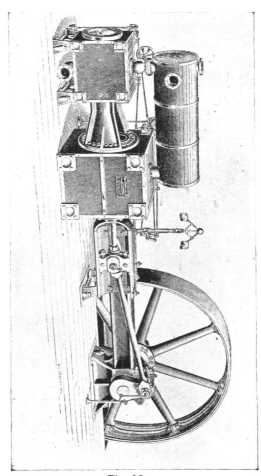

TANDEM COMPOUND REYNOLDS-CORLISS ENGINE.

Fig. 26.

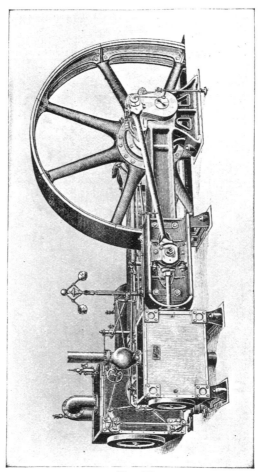

Fig. 27.

CROSS COMPOUND REYNOLDS-CORLISS ENGINE.

ent time.   The releasing mechanism is illustrated in outline in chapter 2, part 2, figure 5, and is styled the "oval arm gear."   Vacuum dash pots are used for closing the steam valves when released at cut off, this style of dash pot being noted for the rapidity of its action at high speeds.

CHAPTER XIV.   THE HARRIS-CORLISS ENGINE.

The engraving, Figure 28, illustrates the Harris-Corliss simple engine.

The releasing gear possesses many desirable and novel features, as will be seen by referring to Figure 29. The use of springs is entirely dispensed with, thereby decreasing the noise so common with other gears, and reducing the wear on pins to a minimum. The engagement of the hook is positive and takes place entirely through the action of gravity, the release being effect by a positive locked edge cam working between the two arms of the hook block lever, and imparts a slight rotative motion to this block, thus unfailingly releasing. The hook contacts have four edges each which may be successively brought into contact as necessitated by wear.

The dash pots of the well known "noiseless" form. They require no piping to conduct away the compressed air, and they adjust themselves readily to variations of load without adjustment. As will be seen by re-

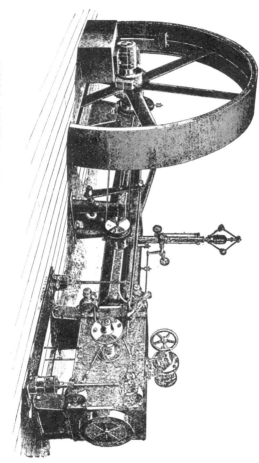

SINGLE CYLINDER HARRIS-CORLISS ENGINE, VALVE GEAR SIDE.

Fig. 28.

ferring to Figure 30 their construction
makes them practically dust proof.

The connecting rods are of the solid
end type with wedge and screw adjustment for the brasses.

The cross-head is of the box pattern,
has large wearing surfaces and a very

Fig. 29.

convenient arrangement for removing
the wrist-pin when taking down the
connecting rod.   The wrist-pin may
be turned to various positions in the
cross-head so as to correct any tendency to wear out of round.

The well known Babbitt and Harris piston is used in all engines built by this company.

All engines over twenty-six inches diameter of cylinder are fitted with two eccentrics for long range cut off.

The governor is of the Porter-Allen type designed to run at a speed of two hundred and twenty-five revolutions a

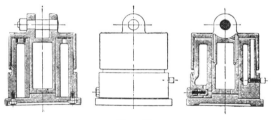

Fig. 30.

minute, with heavy balls and heavy counter-weight, which combination gives it great power and sensitiveness.

CHAPTER XV.   THE PHILADELPHIA
CORLISS ENGINE.

This engine illustrated in Fig.31. Its peculiar features are its massive box pattern frame, and its valve-gear, known as "Gordon's Improved Corliss Valve Gear." It will be seen by referring to Fig. 32, which is an enlarged view of the Gordon valve gear, that the dash pots are cast in one piece with the exhaust valve-stem brackets. They are powerful and noiseless and are so constructed that they discharge no air.

The double ported steam valve is used with this gear, giving a steam line of almost constant pressure up to cut off.

This company also build a "high speed Corliss engine," for electric railway stations and similar work requiring a high rotative speed, as in direct connected engines and dynamos.

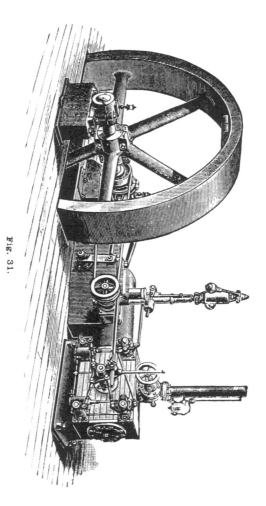

Fig. 31.

PHILADELPHIA CORLISS ENGINE.

Fig. 32.

CHAPTER XVI. THE ECLIPSE-CORLISS ENGINE.

This engine is built in styles and powers to meet the requirements of all classes of modern steam engineering practice.

Fig. 33 is an illustration of a single cylinder girder frame, Eclipse-Corliss engine, and Fig. 34 is a "long range cut-off," tandem compound engine by the same company.

The valve gear is of the usual type of modern design and needs no detailed description, but the valve itself has peculiar features as will be seen in Fig. 35. Instead of being driven by the usual flattened elongation of the valve stem, motion is imparted to the valves by T headed valve stems, and they are held in place by keepers at each end of the valve; they may be removed for inspection without disturbing the valve stems or gear.

The cross head is of the usual box pattern, runs in V guides, and is keyed to the piston rod. It is adjusted by the usual concealed wedge as illustrated in Fig. 36.

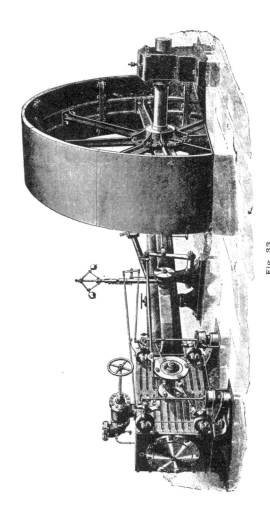

Fig. 33.

SINGLE ECLIPSE CORLISS ENGINE, VALVE GEAR SIDE.

Fig. 54.

TANDEM COMPOUND ECLIPSE CORLISS ENGINE,
(100 H. P.), VALVE SIDE.

A feature of the governor which is shown in Fig. 37, is the "speed adjuster"; by placing the weight at different positions upon the speed lever, considerable variations of speed may be obtained as required.

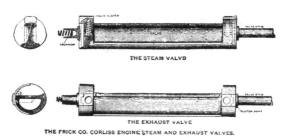

THE STEAM VALVE

THE EXHAUST VALVE

THE FRICK CO. CORLISS ENGINE STEAM AND EXHAUST VALVES.

Fig. 35

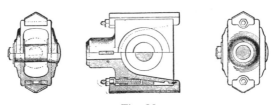

Fig. 36.

Fig. 38 illustrates a tandem compound Eclipse-Corliss engine, driving a double vertical ammonia compressor.

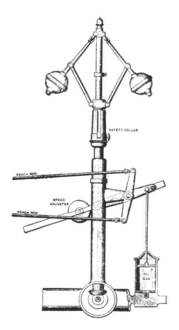

Fig. 37.

Fig. 38.

FRICK-CORLISS ENGINE AND COMPRESSOR.

CHAPTER XVII. THE COLUMBIAN-CORLISS ENGINE.

This engine was produced in honor and commemoration of The Columbian Exposition, and embodies all the improvements in detail and construction demanded by modern conditions of high steam pressure, speed and continuity of service, such as electric light and railway plants and the manufacture of artificial ice.

The Columbian-Corliss engine consists of two main parts—the cylinder and frame. The cylinder in the larger sizes, is bolted directly to the foundation without the interposition of pedestals or legs, and in the smaller sizes the legs are cast on. The pedestals are of box form—in cross section—having two vertical walls of metal for the direct support of each end of the cylinder, at the same time presenting smooth surfaces with no recesses for the lodgement of dirt, thus being easily kept clean.

The frame has the main bearing, with its pedestal, cast upon its outer end, which construction dispenses with useless joints and prevents spring-

ing. Instead of the usual "girder,"
this company have adopted a frame of
box section, supported in the middle
of its length, which is admirably
adapted to withstand complex strains,
and combines the guides, main bearing and seats for the governor and
rocker arm, in one piece. The guides
are of the bored cylindrical style, the
outer ends being tied together by a
heavy ring of metal. Figure 39 is a

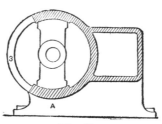

Fig. 39.

cross sectional view through guides, 3
being the ring tying the guides together, and A representing the pedestal under the end of the guides.

The cylinder is fitted with circular
valve bonnets, and has round corners
of large radius on top of each end of
steam chest, which is an improvement
on the square corners and consequent
sharp angle in the steam passages to
the ports. The iron top cast on the
cylinder is one of its peculiar features.

giving it a handsome appearance and doing away with the unsightly warping, shrinking, and swelling of wood lagging.

The steam chest is much larger than usual, and the exhaust chest is separated from the bottom of the cylinder, thereby preventing the cooled exhaust steam from extracting heat from the cylinder walls. The cylinder heads

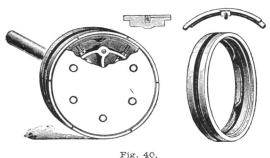

Fig. 40.

are scraped metal to metal, thus making a tight joint without packing.

The piston packing is the well known Babbitt and Harris patent, illustrated in Figure 40. It consists of a chunk ring, with a narrow, sectional, self-adjusting packing ring, automatically expanded by German-silver springs. The chunk ring is provided with the usual centering screws, between it and the spider. When removing this packing from the piston, it is necessary to insert pins—which come with the engine—in the small holes near the circumference of the chunk ring, working them into corresponding holes in the packing ring sections, this will prevent sections of the packing from dropping into the ports in removing or replacing.

The cross-head is the approved box pattern, with removable wrist-pin, and large wearing surfaces.

The connecting rod is of the solid end style with wedge and screw adjustment for taking up the wear of the brasses.

The governor—Figure 41—with which this engine is equipped is extremely simple and wonderfully efficient; the centrifugal force of two balls situated upon the ends of the vertical levers of the bell cranks, is resisted by a spring engaging the inner ends of these bell cranks. By this mechanism the resisting forces can be most accurately adjusted and regulated. It is designed to run at about two hundred revolutions a minute, and owing to its construction the usual dash pot is dispensed with. The safety stop is perfectly automatic, being actuated by gravity in starting the engine.

The valve motion of this engine is fitted with unusually large bearings and pins which is an important fea-

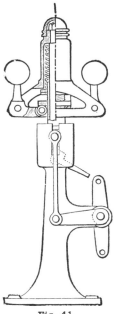

Fig. 41.

ture, for the reason that these joints are usually the first parts to wear loose. The releasing gear is of the oval arm type which has been described; the usual vacuum dash pot is used.

The Heavy Duty Engine, Figure 42, is designed to meet the severe requirements of rolling mills, electric and cable railways. The frame is massive with a bearing practically the entire length of the foundation. The double eccentric valve gear is applied to this style engine; a peculiar feature, adopted by this company, is the absence of the wrist plates. The parallel rods are connected directly to the bell cranks.

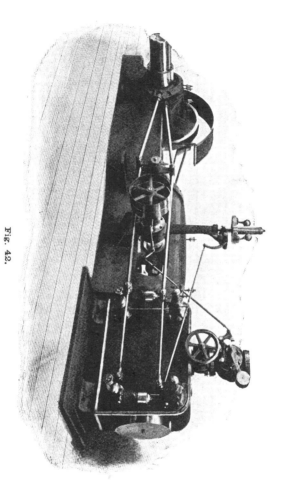

Fig. 42.

HEAVY DUTY COLUMBIAN-CORLISS ENGINE.

CHAPTER XVIII.   THE   FILER   AND STOWELL-CORLISS ENGINE.

This engine is built under the supervision of its designer, Mr. J. H. Vorstman.

The principle features are compactness, rigidity, and simplicity.   All wearing surfaces are made unusually large and provided with improved devices to prevent heating of the bearings.

Cylinders of Corliss engines of large size have been built with ports rather small in proportion to the piston speed, partly because large ports require valves of large diameter, and wide angle of travel, and partly because they increase the clearance.   In the design of this engine these objections have been eliminated, the port areas being of such dimensions that the velocity of the steam is practically the same in all sizes, and the clearance in the valve cavities reduced to a minimum, thereby obtaining high initial and low back pressures.

The frame of the standard pattern is one piece, containing the main bearing

and guides, and rests upon a base or
sole plate of ample dimensions.

The main bearing, Figure 43, is provided with cast iron quarter boxes
lined with babbit metal. The wear
is taken up by heavy adjusting screws
and the quarter box shell is protected
from the wearing in of these screws,
by steel thrust blocks. The upper and

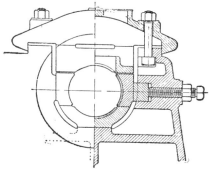

Fig. 43.

lower shells can adjust themselves automatically to the shaft without causing binding or unnecessary friction
and consequent heating. Openings
are provided in the cap, through which
the shaft may be examined by eye and
hand while it is in motion.

For direct connected electric gen-
erators, a special feature is introduced
in the main bearings, whereby the

shaft may be kept in perfect allignment vertically; this is accomplished
by the interposition of a wedge and
screw between the bottom shells and
their seatings.

The cross-head is of a very compact
pattern made of special "semi steel"
which this company use extensively
for details; it is of the box pattern

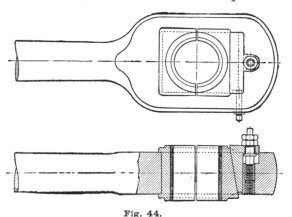

Fig. 44.

with removable wrist pin. The shoes
are turned to fit the guides which are
bored cylindrical.

The connecting rod, Figure 44, is
made with solid ends. It will be noticed that the wedges, instead of being
set vertically in the stub ends, as is
usual with this style, enter the rod at
the side and provide a bearing the full

width and depth of the box, which is very desirable, as this arrangement prevents "wearing in" and consequent springing and heating of the box. The wedge is operated by means of a screw bolt which allows of a very delicate adjustment. A small set screw underneath the rod is added as a safety check. Owing to the disposition of these wedges—the wrist-pin box adjustment being between the wrist-pin and the crank-pin, and the crank-pin box wedge being at the extreme end of rod—the taking up of the wear will leave the distance between the centers of the pins nearly constant, thus correcting any tendency to disarrangement of clearance due to "keying up."

The governor is of the medium speed type with large counter-weight and medium sized balls. A novel safety-stop is introduced which, owing to its peculiar construction, is entirely automatic, and cannot possibly fail to operate should the governor belt run off or break. Figure 45 illustrates this governor so well that further description is unnecessary.

The Heavy Duty "1900" pattern engine, built by this company is illustrated in figure 46. This is a cross compound engine, designed for long

continuous running under heavy
loads, and its construction makes it
well adapted for this purpose.

A complete line of this make of engines are also built, including horizon-

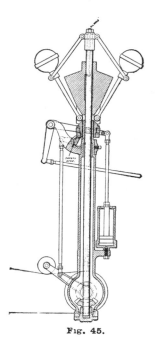

Fig. 45.

tal and vertical engines, either condensing or non-condensing, tandem or
cross-compound, also triple and quadruple expansion engines.

Fig. 46.

CROSS COMPOUND HEAVY DUTY
FILER-STOWELL-CORLISS ENGINE.

CHAPTER XIX.   THE GEO. H. CORLISS ENGINE.

This engine is built by the company which was established in 1849, by Geore H. Corliss, the inventor of the Corliss Engine.

Figure 47 represents a single cylinder engine with two eccentrics and two wrist plates, the latter being of peculiar design. Figure 48 is a view of the crank side of the same engine.

Figure 49 illustrates a four cylinder, triple-expansion engine of 1,000 horsepower, which was built for a New England cotton mill. There are two low pressure cylinders, one being in tandem with the high pressure cylinder, the other one—the left hand in the figure—is in tandem with the intermediate cylinder. This arrangement equalizes the strains and obviates the necessity of using one excessively large low-pressured cylinder.

The heavy duty G. H. Corliss engine is illustrated in Figure 50, which is a cross-compound, "direct connected" engine of 2000 horse-power, which was built for an electric railway in the West. It is fitted with the "long range cut off" on both cylinders.

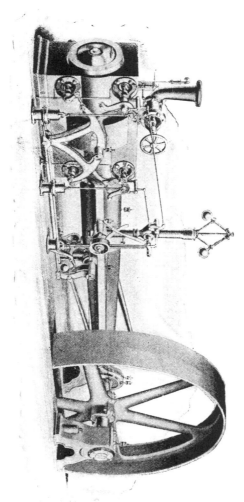

SINGLE CYLINDER GEO. H. CORLISS ENGINE, VALVE GEAR SIDE.

Fig. 47.

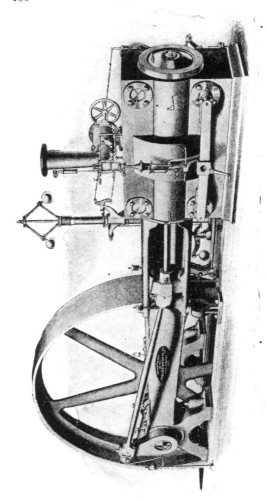

Fig. 48.

SINGLE CYLINDER GEO. H. CORLISS ENGINE, CRANK SIDE.

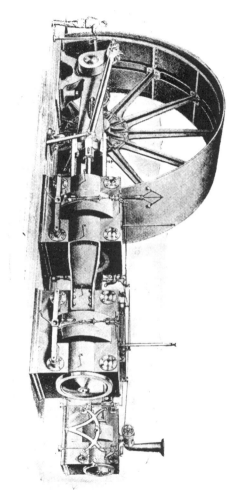

Fig. 49.

FOUR CYLINDER TRIPLE EXPANSION GEO. H. CORLISS ENGINE, (1000 H. P.), BELT DRIVE.

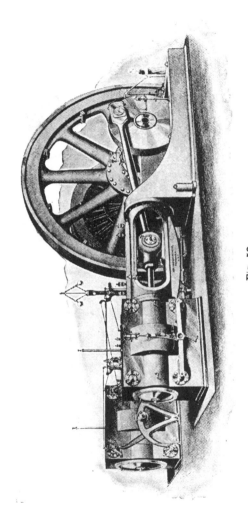

Fig. 50.

CROSS COMPOUND GEO. H. CORLISS ENGINE (2000 H. P.), DIRECT CONNECTED.

CHAPTER XX. THE SIOUX-CORLISS ENGINE.

The standard Sioux-Corliss engine is illustrated in Figure 51. The valve gear is of the approved modern type, a peculiar feature being the disconnecting device—Hart patent—which is one of the latest and best improvements designed to support the reach rod when "unhooked," in order to handle the valve gear with the starting bar. It is composed of two pieces, i.e., a clamp, and a bronze box on the wrist plate pin. The reach rod end is slotted and runs over the box, the latter being ajustable for wear on the pin, a very desirable feature. The clamp is a steel nut with a taper projection, which fits into grooves in the side of the reach rod end, and is fitted with short levers suited to the hand. The general appearance of this device is plainly shown in the illustration.

The governor of this engine is of the high speed type with light fly-balls and heavy counter weight, the latter having a cavity cast in the top intended to receive shot for adjusting the speed to a fraction of a revolution; the

Fig. 51.

SIOUX CORLISS ENGINE.

vertical thrust bearings are all fitted
with hardened steel balls, which pro-
duce an exceedingly light running and
sensitive regulator.   It is provided
with a safety-stop which sets itself automatically as soon as the engine—in
being started—has attained a speed
sufficient to raise the governor a trifle.

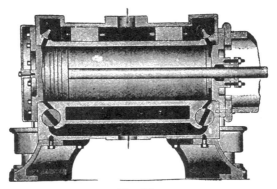

Fig. 52.

By referring to Figure 52, which is a
sectional view of the Sioux-Corliss cylinder, it will be noticed that the exhaust steam passes *through* the exhaust valve instead of over its edge
at one side, also that the valve fills the
valve chamber, thus reducing clearance to a minimum.   The exhaust
chest is separated from the cylinder
walls which is of material benefit **in**
reducing cylinder condensation.

The frame is of the girder type, but of box shape in section and has a heavy pedestal under the end of the guides.

The connecting rod is of the usual solid end pattern with the adjusting wedges for the boxes placed one inside and one outside of the pins which prevents shortening the effective length of the rod in keying up.

The outboard bearing has many desirable features. It is seated upon a sole plate which is provided with a parallel vertical adjustment whereby

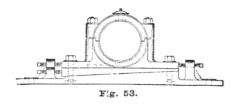

Fig. 53.

the engine shaft may be quickly restored to proper level when thrown out by wear. By removing the tap-bolts which hold the pillow-block to the sole plate, the bearing may be drawn off over the end of the shaft should necessity demand it, making it unnecessary to jack up the shaft as is

usual in a case of this kind, it being
only necessary to take the weight of
the shaft on blocking. Further adjustment is provided for keeping the
shaft square with the center line of the
engine. Figure 53 illustrates these
points.

CHAPTER XXI.   THE VILTER-CORLISS ENGINE.

The Vilter-Corliss engine illustrated in figure 54 is one of the most recent developments of this type of engine with the girder frame.

The cylinder is fitted with circular valve bonnets and circular corners on the top of each end.   The absence of sharp angles in the steam passage gives a free, smooth passage for the entering steam.

The exhaust valves are so constructed that the wearing surfaces come below the valve centers which insures long life of the valves with freedom from leakage. The cylinder is covered with a steel jacket inside of which is placed an approved non-heat conducting filling.

The frame, main bearing and girder for engines up to 18 inches are cast in one piece, the girders in all sizes being of the bored cylindrical style with a pedestal under the outer end.

The usual modern style of valve gear, with oval arm releasing mechanism, and so arranged that it will oper-

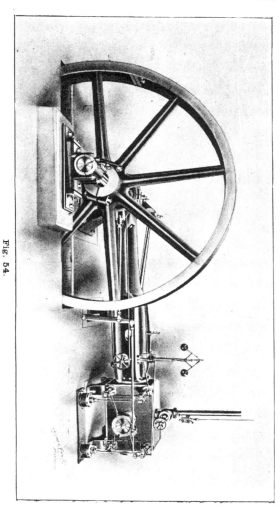

Fig. 54.

VILTER-CORLISS ENGINE.

ate without the use of springs, has been adopted.

The dash pots used on this engine are dust proof, perfectly noiseless, and so constructed that the usual cup-leather packing is dispensed with; the cushion is regulated by turning a small thumb screw as conditions require. They are both mounted upon one sole plate which is bolted to the foundation.

The cross head is of the box pattern with large shoes lined with babbitt metal, and adjusted with a wedge and screw. The wrist pin is a taper fit in the cross-head and is held in place by a nut. The piston rod is either keyed or screwed into the cross head as re quired.

Solid end connecting rods with wedge and screw arrangement for taking up the wrist pin and crank pin brasses, are used.

The outboard bearings are fitted with parallel wedges interposed between the bearings and a heavy sole plate, and are capable of being adjusted both vertically and horizontally without disturbing the anchor bolts, which is a splendid feature.

Simple engines of this pattern are built in sizes from 9x24 inch to 32x54 inch cylinders. Cross and tandem compound engines of this make are also built.

CHAPTER XXII.   THE BATES-CORLISS
ENGINE.

The Bates-Corliss engine, illustrated
in figure 55, differs but slightly if at
all, in general appearance from others
illustrated in this book, but the construction and operation of its valve-
gear are worthy of more than a passing notice.   The use of steel blocks,
springs, hooks, and the usual small
parts have been eliminated in the design of this gear, and an exceedingly
simple "folding device," which accomplishes everything that the hook mechanism does, has been substituted.
The number of parts is noticeably
small, and all joints have pins and
boxes of greatly increased size, thus
the liability to derangement is reduced
to a minimum, and its action rendered
practically noiseless.

The principles governing the adjustments of the ordinary "hook" gear apply equally to this one, as will be readily understood by reference to figure
56, which shows valve gear in full.
W is the wrist plate which gives motion to both steam and exhaust valves.

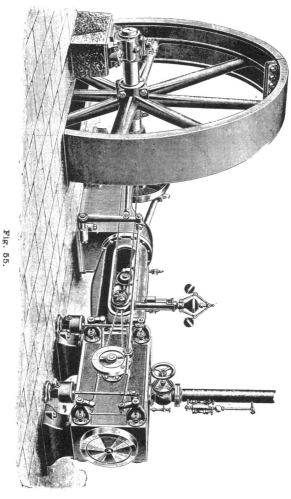

Fig. 55.

BATES-CORLISS ENGINE.

RR are valve rods which operate the
steam valves. LL are connecting
links and are supported by steel pins
II securely fastened in wrist plate. PP
are small steel wrist pins connecting
valve rods RR with links LL. C is a
center line drawn from center of pins
O and I, which indicates the line of
strain between the two points. DD

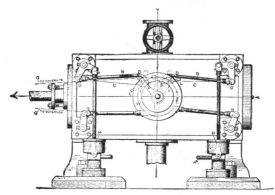

Fig. 56.

are tripping arms moving to and from
each other, varying point of cut-off to
suit load. They are actuated by the
governor through rods GG. HH are
dash pots which instantly close steam
valves as soon as released at wrist
plate. Observe that the center of pin
P on right side which connects link L
to valve rod R is below center line C.

The operation is as follows:—The wrist plate W moving in the direction indicated by the arrow would cause link L to tighten and keep its hold on valve rod R until the end of link L

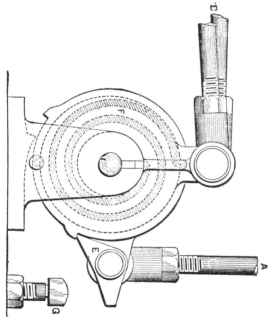

Fig. 57.

comes in contact with roller D at which point the center of pin P is raised above center line C, allowing the dash pot to instantly close the steam valve, the link assuming a sim-

ilar position to that shown on left
hand. When the wrist plate com-
pletes its motion in the direction indicated the left hand link L and rod R
will fold together like that on right
side.

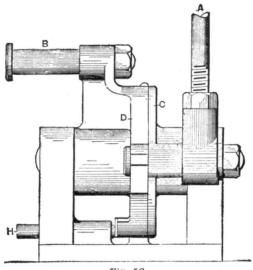

Fig. 58.

The governor is of the weighted fly-
ball pattern and is provided with an
exceedingly efficient, and perfectly automatic safety stop which is ready for
instant action the moment the governor begins to rise, in getting up to
speed.

C and D, Figs. 57 and 58, are independent discs between which is placed spring F connected to the hub of C and rim of D. The tension of this spring is resisted by pawl E on disc C, thus causing discs C and D to work as one. Rod A connects direct to the governor. Rods B connect the tripping device at valve motion. Should any accident befall the governor it would immediately descend until pawl E came in contact with adjustable screw G, disengaging it from disc D, thus allowing the spring F to throw the rods B back to the earliest point of cut-off, shutting off steam and stopping the engine. When the engineer stops his engine and the governor descends, he pushes pin H into a recess in disc D, thus stopping the downward travel of the governor at a point where pawl E will lack just a trifle of being in contact with adjustable screw G. When the engine is started in motion again and the governor rises, the pin H is automatically forced out leaving the automatic stop free to act.

The valves are of generous dimensions and have large wearing surfaces. They are driven from the end, the valve stem being made with a T head,

suitably meshing into the end of the
valve. The use of springs in either
steam or exhaust valves is dispensed
with, their construction rendering
them unnecessary. The valves may
be removed without deranging the
valve gear, which is a decided convenience.

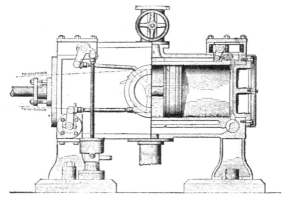

Fig. 59.

These engines are also built with
special admission valves of the flat
slide pattern, driven by the same gear
as the ordinary rotative Corliss valve.
This valve is illustrated in Figure 59.

Engines of this make are built in all
sizes and styles of cross and tandem
compound, vertical and horizontal.

CHAPTER XXIII. THE WATTS CAMP-BELL CORLISS ENGINE.

Figure 60 represents the valve-gear side of a simple Watts-Campbell Corliss Engine arranged for a twin or "pair." The crank-shaft and fly-wheel are made of sufficient strength to transmit double the power of one cylinder, and the end of the shaft, which

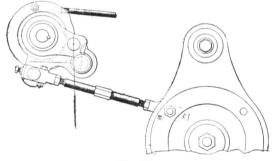

Fig. 61.

projects through the outboard bearing, is provided with a key-way to hold another crank.

By referring to Figure 61, it will be noticed that the releasing device used on these engines differs from the usual form of gear used upon Corliss en-

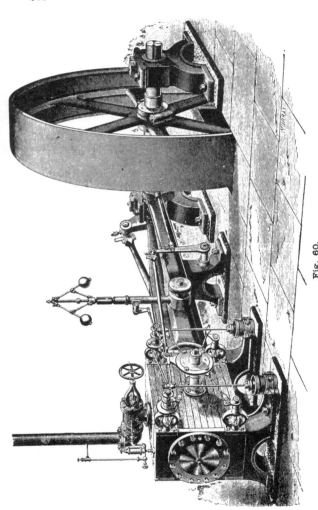

Fig. 60.

THE FISHKILL CORLISS ENGINE.

gines. The releasing arrangement illustrated was devised with a view to eliminating disturbances of the governor at the moment of "knock off," which it very successfully accomplishes. The latch is semi-cylindrical in shape and has a slight rotary motion in hooking on and tripping; the roller upon the end of the latch lever is always in contact with the knock off cam disc, thus avoiding the jar usually sustained when such devices depend upon a blow for the tripping action. The figure illustrates this gear so plainly that a detailed description is unneccessary.

A dash-pot of the usual approved vacuum type is used with this valve gear, and its attachment to the dash pot rod is by means of a ball-and-socket bearing, which permits the dash-pot plunger to turn freely in its bore, thus insuring uniformity of wear and increasing its durability. The ball-and-socket device also compensates for any fault in alignment, should any exist, thus avoiding all danger of binding.

The cross head, illustrated in Figure 62, is of the box type with removable wrist pin and ample bearing surfaces.

The method adopted by the builders
of this engine for adjusting the cross
head in the guides is such that when
the lock nuts are properly screwed up,
the cross head and shoes have the rigidity of one solid piece.   In all engines of this make the piston rod is
keyed into both the cross head and the
piston.

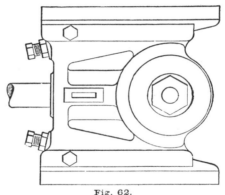

Fig. 62.

The connecting rod is made with the
ordinary strap end, and gib-and-key
adjustment, and is "six cranks" long
or three times the length of stroke of
piston, which is somewhat longer than
the usual practice.

These engines are built in all sizes
from ten inch up to thirty-four inch
cylinders, also cross and tandem compound engines.

CHAPTER XXIV. THE FISHKILL COR-
LISS ENGINE.

The Fishkill Corliss engine is of the
usual design of this type, and is built
with the girder frame of generous dimensions and excellent distribution of
material. Figure 63 is a view of the
valve-gear-side of a simple engine of
this make.

The valves are made of cast iron,
with large wearing surfaces, and may
be removed from their chambers, without disturbing the valve-gear, by taking off the back valve bonnets.

The piston is very strongly built,
and is attached to the piston rod by
a cross key and the end of the rod is
riveted. The weight of the piston is
carried on a junk rink, adjusted by
screws in the spider so that it shall
sustain all the wear, while the spider,
follower and packing rings are kept
central in the cylinder bore. The
packing rings are self adjusting; two
being used in the larger sizes and one
only in the smaller engines. Figure
64 illustrates the design of this piston
thoroughly.

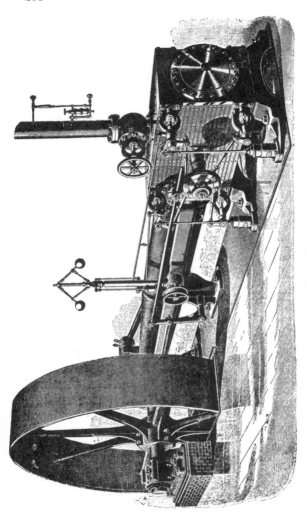

Fig. 63.

THE FISHKILL CORLISS ENGINE.

The cross-head is of the box pattern with removable wrist pin, and is keyed to the pistol rod. The shoes, which have large wearing surfaces are pro-

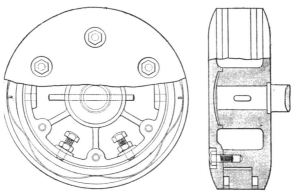

Fig. 64.

vided with a very convenient means of adjustment consisting of taper keys extending across the cross head instead of longitudinally; by the keys the shoes may be quickly and easily removed whenever necessary. See Figure 65.

The connecting rod is of hammered wrought iron, is six cranks long, and is fitted with straps, gibs and, keys in the usual manner.

The principal feature of this engine is its valve-gear, or rather the releas-

ing device, known as Cite's Releasing
Valve-Gear, and is designed to relieve
the governor of the work of actual
tripping, thereby permitting it to
more correctly perform the actual
work of indicating the proper time
when the valves should be released.

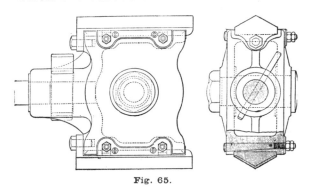

Fig. 65.

The following illustrations show
Cite's Releasing Valve-Gear. Figure
66 is a front elevation, Figure 67 is a
plan, and Figure 68 is a rear elevation
of this device as it appears when engaged, and in the middle of its travel.

In all the figures, A represents the
valve-stem, and B the valve-lever
which is secured to end of valve-stem
by feather and set-screw. C-C' is a
double crank vibrating loosely on a
projection of the bonnet which sup-

ports the valve-stem, and this double-crank is connected by an adjustable link-rod X to the wrist-plate from which it receives its motion. The end of the arm C carries a small rock-shaft

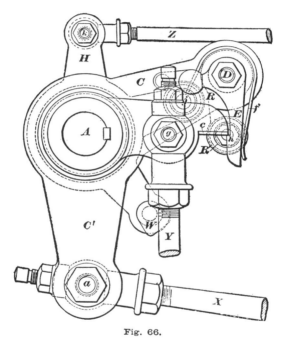

Fig. 66.

D which has a hook E fastened on one end. This hook is provided with a hardened steel catch-plate which engages a similar plate c fastened on the

end of valve-lever B, and the hook is
kept in place by a light spring f.

On the end of rock-shaft D, opposite
the hook E, is fixed a lever F, having a

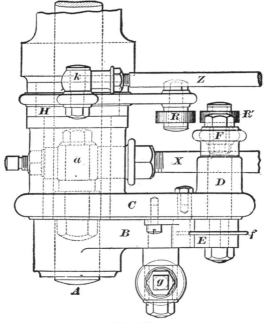

Fig. 67.

pin h on which is mounted a **friction
roller** R. The triple lever HH′ H″ os-
cillates upon a projection of the bonnet which supports the valve-stem;
the arm H is connected by an adjus-

table rod Z to the governor; the arm
H´ has a pin j on which is mounted
a friction rolled R, and on the arm
H″ is mounted an adjustable cam
W (or a friction roller), which is used
for the stop motion.

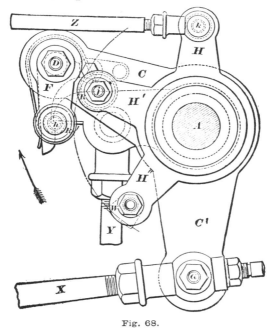

Fig. 68.

By referring to Figure 68, in which
the double crank CC´ is moved by the
wrist-plate in the direction indicated
by the arrow, it will be seen that all

the parts which are connected to the double crank CC' will move around the center of valve-stem A; the side of friction roller R' nearest to the valve-stem will describe an arc of a circle indicated in the figure by a broken line, and when it passes over roller R it will be pushed away from the center of valve-stem A, thereby causing the small rock-shaft D to turn slightly, and at the same time to move the engaging point of hook E far enough to release the valve-lever B, when the dash-pot will act and close the valve.

At the moment of release, the pressure on the triple lever caused by the liberation will be exerted in a radial line from j to A; by the action of the friction rollers R and R' there will be no appreciable strain to turn the triple lever on its axis, and consequently there will be no tendency to disturb the normal action of the governor. As the position of the triple lever is controlled by the governor, any variation in the height of the governor caused by change of load on engine will change the position of point j and of roller R, and so make variations in the times of release of steam valves and in corresponding point of cut-off in steam supply to cylinder.

The action of the Automatic Safety Stop is as follows: When the engine is at its lowest normal speed, and the hook E is at the point of engagement with the valve-lever B, the roller R' comes nearly in contact with the adjustable cam W (or friction roller), which is mounted on arm H" of the triple lever. Now, should the governor belt be broken, or if from any other cause the governor balls should fall below the point corresponding to the lowest normal speed, the triple lever will move in the direction of the arrow, Figure 68; the cam W (or friction roller) will come in the way of the roller R', which will ride on the top of it, thus preventing the hook E from engaging with the end of valve-lever B, and the valve will remain closed. No steam being admitted, the engine will stop.

In connection with the above, a simple attachment is placed on the governor column, by means of which the action of the stop motion may be suspended or made operative at any time by the engineer; and when suspended, the engine can be stopped and started in the unsual way.

INDEX.

PART I.—ERECTING CORLISS ENGINES.

PART II.—ADJUSTING CORLISS VALVES.

APPENDIX TO INDEX.

IMPORTANT!

Please observe the following instructions when ordering.

Remittances should be made by Bank Draft, P. O. or Express Money Orders, or in Registered Letters. Money when otherwise sent, will be at the risk of the sender.

We prepay postage when the price is sent to us in advance.

Books will be sent C. O. D. when $2.00 accompanies the order, but in all cases the customer will have to pay the express charges to him and also for the return of the money to us.

Write your name and address legibly, giving your Post Office, County and State, and be sure to sign your name before mailing.

We take every precaution to insure safe delivery of books, but are not responsible for loss of goods sent by mail.

No books exchanged when sent according to order.

Address all orders plainly to

THE AMERICAN INDUSTRIAL PUBLISHING CO.,
BRIDGEPORT, CONN., U. S. A.

PREFACE.

The preparation of this book has occupied most of the author's spare time for a number of years. Originally the matter was not intended for publication, but the manuscript has grown so large and complete, which consideration, combined with many repeated requests, has induced the writer to publish the matter in book form.

Let every Engineer make his own Indicator book as he proceeds in his study and practice, and it will prove invaluable in after years. The present work has been compiled in this way, from data continually obtained during the author's professional career, extending over a third of a century.

The introduction of algebrical formules have been avoided. These are readily found in the many valuable Mechanical Pocket-Books. The writer has endeavored to discuss the principle and use of the Indicator in as plain common sense words as the subject and the English language will admit of.

Special attention has been given to the requirements of the young progressive student in Steam Engineering. The preparation of the following chapters has been a work of pleasure to the author, and if they prove beneficial to his fellow-workmen, he will be amply repaid.